The Young Charles Darwin

The Young Charles Darwin

Keith Thomson

Yale University Press   New Haven & London

Set in Bulmer type by Binghamton Valley Composition.
Printed in the United States of America.

ISBN: 978-0-300-13608-1

A catalogue record for this book is available from the Library of Congress and
the British Library.

This paper meets the requirements of ANSI/NISO Z39.48-1992 (Permanence of Paper).
It contains 30 percent postconsumer waste (PCW) and is certified
by the Forest Stewardship Council (FSC).

10   9   8   7   6   5   4   3   2   1

Contents

Preface

A whole world of books has been written about Charles Darwin, and rightly so, for he is one of the most intriguing intellectual figures of the nineteenth century, a man whose ideas have continued to dominate society, and particularly science. He published pathbreaking works on subjects as varied as the geology of South America, the causes of coral reef formation, the reproduction of orchids, the origins of humans, the habits of climbing plants, and (in a wonderful little work) the behavior of earthworms. Most of all, of course, he is known for his theory of natural selection, which provided the explanatory mechanism for evolution.

More fascinating even than the shadowy figure of Darwin himself, or the range of his scientific achievements, is the fact that the concept of natural selection—his main idea and such a powerful theory—was generated by one man, working essentially alone. That man has remained inaccessible, hidden behind the familiar caricature of the reclusive but benevolent Victorian genius with white beard and formidable forehead.

Despite all that can be said (and has been said) about Darwin, some glaring questions are not yet fully answered. How and when did Darwin actually come to his revolutionary ideas? How did a quiet and shy twenty-seven-year-old with no apparent credentials as an intellectual, who had moreover been isolated for five years on a small ship (HMS *Beagle*) on a voyage around the world, develop the most dangerous idea of the past two hundred years? The concept of natural selection did not suddenly come to him one day in the bath (as Archimedes is supposed to have discovered the phenomenon of specific gravity).

On what intellectual foundation did he create such a theory out of a mass of information and a complex of rival theories, and in a context of intellectual and social turmoil?

The core of Darwin's idea is elegant in its simplicity and daunting in the range of its consequences. He developed it primarily over a period of five years, from 1837 to 1842, but had been prepared for the subject by his studies at Edinburgh and Cambridge universities. He developed much of the necessary empirical data and honed his analytical skills while on board HMS *Beagle*. He continued to refine his theory until it was finally published in 1859 as *On the Origin of Species by Means of Natural Selection, or the Preservation of Favoured Races in the Struggle for Life,* and in a sense he never finished modifying it.

The purpose of this book is not to retell the story of Charles Darwin's life, which has already been done superbly well by (leading all others) Janet Browne, Adrian Desmond, and James Moore. Rather, it is to inquire into the range of influences and ideas, the mentors and the rivals, and the formal and informal education that shaped Darwin's thoughts on the subject of evolution. He did not produce his theory of natural selection out of the clear blue sky. Many others before him had started down the same path and faltered. What special qualities of mind, what unique experiences, and what intellectual debts did Darwin bring to bear on what, in his *On the Origin of Species,* he called the "mystery of mysteries—the first appearance of new beings on this earth"?

A huge number of influences may have affected Darwin's work. Much has been written elsewhere on the social, political, and religious context of Darwin's life and its contribution to his thinking. Coauthors James Moore and Adrian Desmond are among the most notable explorers of that world. I try here to concentrate on the science itself and to tell the story of his changing scientific ideas, his teachers, his colleagues, and his firsthand experiences as a naturalist. This book ends where many others are just getting going—at the point that Darwin had a working version of his theory in 1842 and a fuller one in 1844.

The development of Darwin's evolutionary ideas was not a simple straight line, and it happened while Darwin was engrossed in other subjects, with a workload and an intellectual load that would have been too much for many a person. There was a myriad of facts and ideas he had to sort through, retain or discard. He followed many blind alleys. He had to jettison many old-fashioned beliefs. Darwin started off with only a general idea of the direction in which he was heading.

How he actually got to his revolutionary theory is one of the most fascinating stories in science, as we see a brilliant mind probing into unknown and even dangerous territory. The birth and maturing of his theory are striking both for the way they reveal his genius and for demonstrating that he was very much a mortal man, with human foibles and weaknesses. Little, if any, of the raw science that underlies Darwin's theory was discovered from his own labors. Like Newton, he saw further because he stood on the shoulders of giants and, like Newton, he was himself a giant. Darwin's great talent lay in his ability to sift through a mass of data and ideas to find the kernels of new truths and their essential interdependencies. The course that he followed to carve out his idea was—like any human endeavor—tortuous and complex.

A Note on Names

In any biographical work there exists the problem of how to refer to the subject—by the given name or surname. As its family tree shows, the Darwin family has been quite sparing in the bestowing of first names: Erasmus, Robert, and Charles appear in several generations of the men. Not only did the famous Charles Robert Darwin have an uncle named Charles, for example, his grandfather, another uncle, and his brother were all named Erasmus. This creates a dilemma in being completely clear to which Darwin one is referring. I have been unwilling to refer to the subject of this book as Charles, although others authors have done so; here, unless clarity absolutely requires more information, I have referred to him simply as Darwin. Members of previous generations are given their full names. Darwin's siblings are referred to, somewhat familiarly, by their first names, as is his wife, Emma. I have not used the nicknames they had for each other (Charles—Bobby; Erasmus—Ras; Caroline—Granny; Emma—Titty).

A Note on Spelling and Punctuation

I have tried here to let the main characters—including, of course, Darwin himself—speak for themselves. In Darwin's case particularly, that means that all quotations are full of eccentric punctuation and haphazard spelling. It would be impossibly disruptive to put "[*sic*]" next to each example. Therefore, I have left things just as Darwin wrote them (for example, "muscles" instead of "mussels"). This preserves the immediacy (and charm) of his diary and notebook entries, while driving copy editors crazy.

Acknowledgments

In winding up the writing of any of my books, my grateful thoughts always go first to the libraries that, with their rich resources and welcoming staffs, have made the work possible. I must particularly acknowledge here the help I have received for many years from the Cambridge University Library (Peter Gautrey and Adam Perkins), the Bodleian Library of the University of Oxford and the library of the Oxford University Museum, the American Philosophical Society (Martin Levitt, Roy Goodman, Charles Greifenstein, Valerie-Ann Lutz, Earl Spamer), and the Ewell Sale Stewart Library of the Academy of Natural Sciences of Philadelphia (Elaine Matthias, Robert McCracken Peck). I am grateful to Mrs. Ursula Mommens for permission to quote from a manuscript list of Darwin's reading at Edinburgh that she has deposited in the Darwin Archive at Cambridge University, and to the Syndics of Cambridge University Library for permission to quote extensively from Darwin's correspondence and notebooks.

Any student of Darwin also owes a very great deal to the stars of Darwinian scholarship whose work has so fully illuminated his life and work. They are too numerous to mention by name; their citations in the endnotes speak for themselves. However, it is necessary to single out here the enormously erudite scholarship that has gone into producing the definitive edition of Darwin's *Correspondence*, now in its sixteenth volume, begun by Frederick Burkhardt and Sydney Smith for Cambridge University Press. I am grateful to Stephen Keynes and Randal Keynes for their friendship and an entrée to the Darwin family. Kevin Padian and Linda Price Thomson read the manuscript with their usual critical eye. Russell Breish, MD, offered valuable comments on the literature about Darwin's illnesses.

Finally, I am deeply grateful to my editor, Jean Thomson Black (no relation), and to Matthew Laird at Yale University Press, and to my agents, Felicity Bryan and George Lucas, for their constant support.

PART ONE

Beginnings

ONE

Falmouth

On Sunday, October 2, 1836, in gaps between the gray squalls roaring in off the Atlantic, an observer on Pendennis Point or St. Anthony Head might have spotted a small ship quartering across the seas and taking on a great deal of water as it headed for the relative calm of Falmouth, Devon. His Majesty's surveying ship HMS *Beagle,* a ninety-foot converted brig, slipping unannounced into Falmouth Harbor, was returning home after a voyage around the world lasting four years and 278 days. On board were Lieutenant Robert FitzRoy, her captain, some sixty officers and sailors, and a few landsmen. After so long away, FitzRoy wanted to get letters off to the Admiralty in London announcing his arrival—and no doubt also he was eager to communicate with Mary O'Brien, soon to be his fiancée, a secret kept from everyone, even his closest confidant on the vessel, the ship's naturalist, Charles Robert Darwin.

The voyage of HMS *Beagle* was fated to go down in history as one of the epic journeys of both hydrographic science and natural history. It would become associated with Darwin's name far more than FitzRoy's, especially when, in 1839, Darwin first published the book that we now know as *The Voyage of the Beagle.*

On this date in 1836, Darwin was thoroughly sick of the *Beagle* and her officers. He hated the sea and everything about it: the cold, the damp, the smells, the close quarters, the mindless routines—and above all the nausea of seasickness. After such a long voyage, the last days of which were spent in coping with the heavy weather one typically encounters when crossing the Bay of Biscay, it was natural that he and everyone on board would be homesick for England and eagerly crowding the sides of the little ship for a glimpse of the green hills of home. But on this stormy and cheerless day, much to his surprise, Darwin, who had been aching for this moment, found that "the first sight of the shores of England inspired me with no warmer feelings, than if it had been a miserable Portugeese settlement."[1]

The *Beagle* had stopped at Falmouth briefly en route to its destination of Plymouth, where FitzRoy was expected to make his last measurement of longitude from exactly the same spot from which they had started. One of the principal purposes of the *Beagle* voyage had been to make a single continuous suite of measurements of longitude around the 360-degree circle of the globe. This had never been done before by a single ship and the same set of observers. It required using chronometers that could be kept running precisely on time for nearly five years. The *Beagle* had started with twenty-two of the best chronometers that could be bought or borrowed; at least three of them seemed still to be running well when they hove into Falmouth.

Everyone on board was anxious to see what the result would be: how closely had they come to that complete circle? The results would matter a lot for the accuracy of the geographical coordinates that they had fixed for sites around the world. If they had been accurate, future sailors would have confidence in their maps and charts. Right at the beginning of the voyage, for example, FitzRoy had discovered that the available charts for South America had Bahia de Todos los Santos (now San Salvador), Brazil, off by some four nautical miles. That could be a disastrously wide gap for anyone navigating by dead reckoning on a dark, stormy day like this one.

FitzRoy set off again from Falmouth on October 4 for the short jog east along the Devon coast to Plymouth, where he was to pay off the crew. But by then Darwin was missing. Unconscionably, Darwin had jumped ship. On the night of the 2nd, with the coast battered by more fierce storms, he took the night mail stage, heading for Shrewsbury and home.

Darwin's role on the *Beagle* had been complicated right from the beginning. He was only a supernumerary crew member, essentially a guest of FitzRoy, taken along to provide expertise in natural history for the *Beagle*'s epic circumnavigation of the world. Through all those years of voyaging, he had lived at close quarters with FitzRoy and the other officers and crew of the ship, sharing triumphs and pleasures, losses and setbacks, joy and misery alike. He was bound to those men like a brother, but still, even if it was bad form, he could not wait—did not wait—to get away.

In justification of Darwin's behavior, in the storms that the *Beagle* had just come through he had been seasick yet again. Perhaps uniquely in the annals of exploration, Darwin never "got his sea legs"; he was just as seasick in October 1836 as he had been in December 1831—when the ship was still in harbor. Once he had seen the terra firma of Devonshire, gray and cheerless as it appeared, the prospect of even a few more hours on board, let along the final leg to Plymouth, must have seemed like a sentence of death.

More than that, Darwin, who had started the voyage with only vague prospects of what his future life might look like, was a man in a hurry. He was now twenty-seven. His contemporaries were established in careers; most were married with families. He, on the other hand, had not yet really started out. He was in a hurry because, on the long journey north from the Cape of Good Hope, with a last-minute detour to South America, he had finally seen exactly what his future could hold. Intense, cerebral, still maturing, shy in company, confident in his abilities, Darwin had discovered during the voyage that he could frame important ideas in the fields of both geology and biology. In his notebooks he had the materials for no fewer than three books, and possibly more. Having been in isolation for so long, he was desperate to be in the company of scientific men who would either confirm or deny his ideas. While many of the officers of the *Beagle*, young as they were, were thinking that this voyage might mark the end of their careers, for Darwin everything was now beginning.

Robert FitzRoy also had a very keen sense of what the *Beagle* voyage had accomplished. He was not going to be satisfied with quietly putting in at Plymouth and ending the voyage there. He requested permission to take the *Beagle* round into the Thames Estuary, to the Greenwich Meridian itself. There, he rightly assumed, he, the ship, and its officers and men would get the heroes' welcome they deserved.

Darwin really should have stayed with the ship at least until the eagerly awaited chronometric measurements were taken at Plymouth; he should have stayed to help share in FitzRoy's triumph—it turned out that they were only some five seconds off a full 360-degree circle: a simply phenomenal performance. But after nearly five years of intense experiences in natural history, Darwin's time for adventure was over; the man who had started the voyage at twenty-two with a vague prospect of becoming a scholarly country parson was now committed to three projects that would change the shape of natural science for ever.

The first project was to write up his journals from the voyage as a book of scientific travels. Right from the beginning he had been sending home portions for his family to read and comment on. FitzRoy had invited him to contribute that diary to the formal account of the voyage (and of its predecessor, an aborted survey of South America between 1826 and 1830). But Darwin knew that what he really wanted was to write his own book. In the end, his story of the voyage became the third volume of the four-volume official *Narrative of the Surveying Voyages of HMS Adventure and HMS Beagle* and was soon reprinted under the title by which we know it today, *The Voyage of the Beagle.*

Darwin's second projected work was even more ambitious: an account of the geology of South America. This would have been a mammoth task for any geologist. Darwin, who had seen the plains of Patagonia, followed the Santa Cruz River inland almost to its source, rounded Cape Horn, and climbed the Andes, completed his *Geological Observations on South America* (in two volumes) in 1846 (having initially written a number of scientific papers on the subject, the first of which had been published by the Geological Society of London before the *Beagle* made landfall).

His third foray into natural science would be to write up his revolutionary ideas about the origins of coral atolls, those strange ring-shaped structures found in tropical seas. On this subject, his ideas directly contradicted the views of the greatest geologist of the age (and of the ages), Charles Lyell. But within months of Darwin's return, that great man had accepted them as being superior to his own explanations. Bursting with all these—and many other—ideas, Darwin simply had to get off the ship. There was not another moment to waste.

The fact that within a very short period Darwin would accomplish everything that he had in mind when he left the ship presents us with a fascinating puzzle. He was evidently an intellectual of the first order. If he had never contributed to the field of evolution, all the other works of his long lifetime would have secured him a place as one of the world's greatest natural scientists. He was recognized as an important figure within a very few months of his return to England. But if, in 1831, when the *Beagle* left England, he had been—as everyone said of him—merely an amiable young amateur naturalist, how could he have emerged from five years of isolation on board as someone capable of making such significant intellectual contributions to geology and natural science? He had been expected only to collect specimens and make extensive field notes. How did it come about that, at some very early point in the voyage, the intended clergyman started to think seriously about some of the most philosophically and empirically difficult issues in all of science? Perhaps the most interesting question of all is: when exactly did he start on a fourth—and much more difficult—project: to think seriously about what we now call "evolution"? Was that also one of the revolutions that had its origins in the voyage?

But first he had to get home. Sadly, the first glimpses of those English landscapes that he had been pining for so keenly continued to disappoint. Black and gray continued to predominate, rather than the expected lush green, as the mail headed north. The following day, as the coach rattled further across the West Country, the weather started to improve and, with it, his spirits rose. He started to miss his shipmates after all. He would have had to be quite inhuman not to.

He arrived in Shrewsbury in the early morning and, pale and thin, walked in on the family as they started breakfast. His sister Caroline wrote to her cousin Sarah Wedgwood, "We heard nothing of him till this morning. . . . We have had the very happiest morning—poor Charles so full of affection & delight at seeing my father looking so well & being with us all again—his hatred of the sea is as intense as even I can wish."[2]

Two days later, he wrote to FitzRoy apologizing for his behavior. The letter is quite revealing of the bonds that had grown between the two young men.

> I am thoroughly ashamed of myself; in what a dead and half alive state, I spent the few last days on board, my only excuse is, that certainly I was not quite well.—The first day in the mail tired me but as I drew nearer to Shrewsbury everything looked more beautiful and cheerful—In passing Gloucestershire & Worcestershire I wished much for you to admire the fields woods & orchards.—The stupid people on the coach did not seem to think the fields one bit greener than usual but I am sure, we would have thoroughly agreed, that the wide world does not contain so happy a prospect as the rich cultivated land of England. . . . I thought when I began this letter I would convince you what a steady & sober frame of mind I was in. But I find I am writing most precious nonsense. Two or three of our laborers yesterday immediately set to work, and got most excessively drunk in honour of the arrival of Master Charles.— Who then shall gainsay if Master Charles himself chooses to make himself a fool. Good bye—God bless you—I hope you are happy, but much wiser than your most sincere but unworthy Philos.[3]

All of that was tactless in the extreme, as FitzRoy still had got the *Beagle* no further than Plymouth and was negotiating to go on to London for a triumphant arrival at Greenwich. It would be many weeks before FitzRoy would be home. Meanwhile, he took the ship on from Plymouth to Portsmouth, where everyone had to languish until further orders came from London. (FitzRoy's letters from Falmouth had at first been put aside unread.)

When the *Beagle* finally arrived at Greenwich, it was indeed to a hero's welcome. The exploits of the little ship in circumnavigating the world were already widely talked about. The great and good came aboard to show their respects and see the tiny ship for themselves. In one famous episode, the astronomer royal arrived with his wife. Instead of being shown to the accommodation ladder with "respectable persons," they were directed by a sailor to the gangway, which meant clambering up a set of cleats on the ship's side with the aid of two ropes for handholds. Which they gamely did, climbing over the gunwale onto the deck—not an easy thing to do in a decent suit, let alone a full skirt. "Well, sir, they did not look respectable" was the sailor's explanation.[4]

Darwin did eventually travel to Greenwich to visit his old shipmates for one last time before the ship was moved off to Woolwich Dockyard and the crew discharged. In the intervening weeks, he had discovered that not only was the voyage of the ship famous, so was he. All along, he had been sending letters and specimens to his old tutor at Cambridge, the Reverend Professor John Stevens Henslow. In these letters he had poured forth his natural history observations. Henslow had them published by the Cambridge Philosophical Society. His discoveries of plants, animals, and fossils and his descriptions of geology and natural landscapes were important. Whatever the strength of his secret ambitions to make a name in science, his confidence in his own potential had already been confirmed.

Even so, on the face of it, little could be more implausible than that this particular young naturalist should have developed the idea—so logical, so dangerous—that has uniquely dominated biological science, and much else, for 150 years.

TWO

Antecedents

Motherless from the age of eight, Charles Darwin was evidently an unusual child: imaginative but an inattentive student, impatient with formal learning but an avid reader, undirected but passionate about his scientific interests (natural history and chemistry), spoiled by his older sisters, adored by his younger sister, Catherine, whom he liked to order around, and fiercely attached to his brother, five years his senior. In many respects he was rather asocial. Instead of playing with friends, he took long, solitary, pensive walks. Once he was so deep in thought as he walked along that he fell off a bank. Of his time at Shrewsbury School the only serious interests that he mentioned were self-directed. He recalled spending many hours alone sitting in a deep window reading Shakespeare. As he grew into his teens, he became unusually tall, athletic, quite handsome with a large forehead, and already possessed of unusual qualities of concentration and intelligence. He was later remembered as appearing to be two different people. To an inner circle of family and friends, he was the affable and sporting "good chap," cheerful and good company. Otherwise, he was withdrawn, aloof, and disdainful of others if they did not meet his standards.

Darwin's hobbies were many, but all focused around nature: "When 9 or 10 I distinctly recollect the desire I had of being able to know something of every pebble in front of the Hall door—it was my earliest—only geological aspiration at that time. . . . I do not remember any mental pursuits excepting those of collecting stones &c.—gardening & about this time often going with my father in his carriage, telling him of my lessons, & seeing game & other wild birds, which was a great delight to me.—I was borne a naturalist."[1]

Darwin, as is well known, was born on February 12, 1809, thus providing an answer to the Marxist (Groucho, not Karl) riddle: "Who was born on Lincoln's birthday?" His father was Robert Waring Darwin (1766–1848) and his mother was Susannah Wedgwood (1765–1817). Darwin was next to the youngest of six children, and he may have resented losing his status as the pampered baby of the family when his younger sister, Catherine, was born. Darwin's only brother, Erasmus, on whom he doted, was named after their illustrious grandfather. Erasmus seems to have been marked both as a child and a man as charming, intelligent, and sociable but weak. All the Darwin children were born into a clan in which *family* and *achievement* were paramount, and characters were richly drawn and larger than life. Indolence, lack of discipline, and passivity were not tolerated.

Although Robert Waring Darwin was a powerful element in Darwin's life, the great intellectual presence in the background was his paternal grandfather, Erasmus Darwin (1731–1802), intellectual, noted doctor and prolific author.[2] Flamboyant and iconoclastic, nonconformist in religion and unconventional in philosophy, Erasmus Darwin was more a product of the age of Rousseau and the French Revolution than of Adam Smith and David Hume. Brought up in Unitarianism, which he dismissively termed "a feather bed to catch a falling Christian," he would have been happiest being termed a humanist, one who accepted the existence of God and Christ's humanity but denied the necessity of Christ's divinity.[3]

Erasmus Darwin's intellectual reach was remarkable, and his fame was possibly no less than that other son of Lichfield, Dr. Johnson (of dictionary fame), with whom he constantly felt himself in competition. From his Midlands base, he became a member of the famous Lunar Society of scientists, philosophers, and entrepreneurs that included James Watt, Joseph Priestley, Matthew Boulton, and Josiah Wedgwood. They

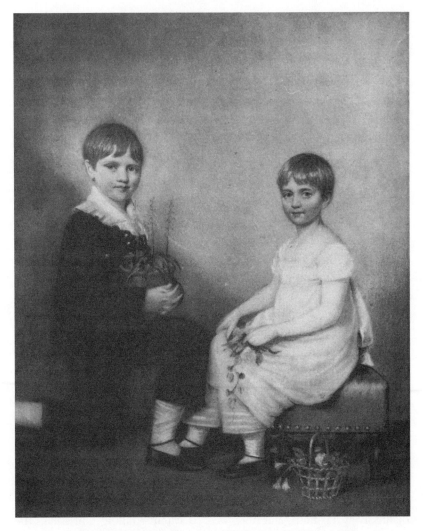

1. Charles Darwin aged seven, with his sister Catherine, 1816.
(Courtesy of Cambridge University Library [MSS.DAR 225:109].)

met monthly; the "Lunar" name is supposed to refer to them holding
meetings on nights when the moon would guide their way home (a vain
hope under the frequently cloudy skies of England).[4] Like his Lunar
Society colleagues, Erasmus Darwin was a strong believer in progress
and the mastery of one's own fate. In the industrial English heartland,

rapidly growing in economic and political power, the more repressive effects of English society and the class system were less strongly felt.

In this world where most things were possible to someone of drive and intelligence, the Lunar Society members were intensely practical men. Erasmus Darwin was himself an inventor of great imagination.[5] Among the devices that he designed were an improved oil lamp, a copying machine, a speaking machine, an alarm clock, a steam-powered carriage, a water pump, and a steam turbine. Not all of his ideas were successes. His idea for improving the ride of a carriage, using springy steel for the wheel spokes, tended to result in the passengers being thrown out.

Above all, Erasmus Darwin wished to be remembered as a philosopher-poet. His major works were literary. In his *Zoonomia* of 1794, he sought to develop a unifying theory of nature. "The purport of the following pages is an endeavour to reduce the facts belonging to ANIMAL LIFE into classes, orders, genera, and species; and, by comparing them with each other, to unravel the theory of diseases. . . . A theory founded upon nature, that should bind together the scattered facts of medical knowledge, and converge into one point of view the laws of organic life, would thus on many accounts contribute to the interest of society. It would capacitate men of moderate abilities to practice the art of healing with real advantage to the public; it would enable every one of literary acquirements to distinguish the genuine disciples of medicine from those of boastful effrontery, or of wily address; and would teach mankind in some important situations the *knowledge of themselves.*"[6]

Zoonomia is a long, rambling work, difficult to read and perhaps interesting only as an index to medical knowledge of the time—except for one crucial matter. It is unquestionably the first full-length treatise to include a theory of the transmutation of species (evolution). In *Zoonomia* Erasmus Darwin argued that all living forms are related to each other in patterns of relationship by descent. In later works such as *The Temple of Nature* (1803), he proposed that the first life arose, naturally, chemically, out of nonlife, according to natural laws and processes.

Of course, there was no direct evidence for, and a great deal of cultural tradition against, an evolutionary view of life. No causal mechanism, if such had indeed existed, was known, although none other than Linnaeus himself had suggested that new species of plants might arise through hybridization. The turn of the nineteenth century was also just the wrong time to be proposing revolutionary ideas of change: the French Revolution had been watched with horror from the northern shores of the English Channel, and the British establishment wanted none of it. Erasmus Darwin's ideas probably did not reach the audience that they might otherwise have enjoyed.

Even in the present day, the evolution (transmutation) of species one from another according to natural, material principles and processes, over vast eons of time, is a matter of great contention. After Erasmus Darwin, evolution quickly became a prime focus of the age-old debate between the roles of science and religion as different—sometimes rival, sometimes complementary—worldviews and systems for "knowing." It was a prime example of the aphorism "Science does not make it impossible to believe in God; it just makes it possible not to believe in God."

Erasmus Darwin's view of transmutation of species preceded the ideas of his grandson by sixty years. But it was not a new subject; scholars had argued about it for centuries, mostly to deny its possibility. The Bible, after all, said that God had created all living things at a single point in time. Since God's work was always perfect, no further changes would have been necessary. Like many of his contemporaries (Georges Cuvier in France, for example), Erasmus Darwin had a fascination for fossils, and these told a different story. Other species had once lived on earth, and they had been different from those living today. Where had they come from? A religious view would have been that God had made them, too. But the result of the Enlightenment was to embolden scholars to seek for lawful processes in nature, based entirely on material—not supranatural—processes and properties. The responsibility of the philosopher was, then, to imagine the unimaginable.

Erasmus Darwin's theory of transmutation grew out of his analysis, as a medical man, of early embryology, which was then very poorly understood. If each individual life begins as a single cell or tiny formless mass of cells ("a living filament with certain capabilities of irritation,

sensation, volition, and association"), then possibly all life on earth began in the same way. "In some this filament in its advance to maturity has
acquired hands and fingers, with a fine sense of touch, as in mankind. In
others it has acquired claws or talons, as in tygers and eagles. In others,
toes with an intervening web, or membrane, as in seals and geese. In others it has acquired cloven hoofs, as in cows and swine; and whole hoofs
in others, as in the horse. While in the bird kind this original living filament has put forth wings instead of arms or legs, and feathers instead
of hair."[7]

Erasmus Darwin's whole idea depended on the capacity of various
external conditions to change the developmental trajectory of this "filament." He thought that there were several classes of such modifying
factors: selective breeding (of which the prime example would be animal and plant husbandry), variation by mutation and "monsters," and
"exercise to gratify" three basic drives: lust, hunger, and security. This
"exercise" involved a healthy dose of volition, the will of the organism
to change (or to stay the same as its parents), and the powerful reinforc-
ing effects of habitual use and disuse of features.

His conclusions were daring but clothed in pious terms: "In the
great length of time, since the earth began to exist, perhaps millions of
ages before the commencement of the history of mankind, would it be
too bold to imagine, that all warm-blooded animals have arisen from
one living filament, which THE GREAT FIRST CAUSE endued with animal-
ity, with the power of acquiring new parts, attended with new propensi-
ties, directed by irritations, sensations, volitions, and associations; and
thus possessing the faculty of continuing to improve by its own inher-
ent activity, and of delivering down those improvements by generation
to its posterity, world without end!"[8]

At the turn of the nineteenth century, a French zoologist, Jean-
Baptiste de Lamarck, developed a theory of transmutation, the model
for which was Erasmus Darwin's, and his ideas continued to have currency even after the theory of evolution by natural selection had been
proposed by Erasmus Darwin's grandson.

Ugly, smallpox marked, obese, stammering, lame from a childhood
injury, but with a brilliant mind, Erasmus Darwin charmed everyone he

met. "Florid health, and the earnest of good humor, a sunny smile, on entering a room, and on first accosting his friends, rendered, in his youth, that exterior agreeable, to which beauty and symmetry had not been propitious."[9] In contrast to many of his children and grandchildren, Erasmus Darwin was always blessed with good health, except for hemorrhoids and gout in his later years, which he tried to control by means of diet. As a professional he was a decided success; by 1772 his medical practice was already bringing in a profit of more than £1,000 a year (at least £100,000 today).

Even when his views were unpopular—he was a republican when it was not always expedient to be so—and although his manners were often eccentric, Darwin dominated the town of Lichfield. His friends were legion and they almost universally forgave him his oddities and peccadilloes. Behind the bonhomie, however, Erasmus Darwin was a dark, stern man, sarcastic, with a violent anger. His children often pleased him less than did his friends. He disliked open shows of affection or any display of emotion in a man. He tended to browbeat people, even in his medical practice.

A life of the old doctor published in 1804 by Anna Seward ("the Swan of Litchfield"), who had been Erasmus Darwin's disciple, friend, and possibly more, caused something of a scandal.[10] The work frankly celebrated his intellect and exposed his venality. She cheerfully, even enthusiastically, demonstrated that Erasmus Darwin was a man of large appetites—for ideas, for food, and for the ladies. One of her more shocking inclusions from their long, on-again, off-again relationship was a letter from Erasmus Darwin (Mr. Persian Snow) to Seward (Dear Pussey) full of what her readers found to be the most egregious double entendres. In this work, along with a critical (perhaps, more accurately, an uncritical) reading of Erasmus Darwin's writings, she also discussed the characters of his children and vividly described the weaknesses that led to the suicide of his son, also named Erasmus.[11]

Erasmus Darwin's first wife, Mary Howard, whom he married in 1757, was the daughter of a Lichfield solicitor. After bearing five children, she died at the age of thirty of acute alcoholism, probably exacerbated by overdoses of opium prescribed by her husband. Three sons,

Charles, Erasmus, and Robert, survived. A daughter, Elizabeth, and a fourth son, William Alvey, died in infancy. Charles (1758–78), the first child, was his father's favorite and the great hope of the family. Like his father, he stammered badly, but at medical school in Edinburgh he was a brilliant student and by all accounts a delightful person. He was interested in mechanics like his father and also wrote poetry. Everyone expected great things from him; perhaps he might have become an intellectual in his father's mold; at the very least, he was destined to be a notable doctor. But while still a student at the age of nineteen, he contracted an infection from a cut during an autopsy and died of acute septicemia. On his tombstone was written: "Possessed of uncommon abilities and activity, he had acquired knowledge in every department of medicinal and philosophical science, much beyond his years."

The second son, Erasmus (b. 1759), wanted to enter the church. Erasmus Darwin discouraged this, so he began a career in the law, establishing a solicitor's practice in Lichfield. But he became a strange and withdrawn man, always unable to meet the weight of his father's expectations. Anna Seward reported: "There was a want of energy in his character, and too extreme a delicacy of feeling on the occurrence of every thing that was in the slightest degree repulsive."[12] Very little was said about him in later years within the Darwinian circle, except that his father was "often vexed at his retiring nature and at his not more fully displaying his great talents." His condition has also been attributed to the fact that he lost his mother when he was eleven years old and his older brother Charles when he was nineteen.[13] He may not have been helped by the arrival of eight half siblings in his short lifetime. He committed suicide, a bachelor, at the age of forty. Old Erasmus Darwin was devastated.

Eleven years after Mary Howard died, Erasmus Darwin married Elizabeth Pole, a widow of great beauty and independence of spirit who was the illegitimate daughter of the Earl of Portmore. He moved his practice to Derby with great success. It was at Derby that he joined with the members of the Lunar Society, and his life and writings took on a new brilliance.

Elizabeth Pole assumed the upbringing of fifteen-year-old Robert Waring Darwin, the surviving son, and then bore seven more children

(Edward, Frances, Emma, Francis, John, Henry, and Harriet). Meanwhile, in between marriages, Erasmus Darwin had fathered two daughters on Mary Parker, the governess of his older children.[14] In typical Erasmus Darwin fashion, while matters of class prevented him from marrying Miss Parker, the two girls were fully accepted into his family and his circle of friends.

Robert Waring Darwin—the father of Charles Darwin—became a prominent provincial doctor. He was even more successful than his father as a speculator and an astute moneylender, but he was interested in neither literature nor philosophy. With none of the intellectual inclinations or poetic flights of Erasmus Darwin, Robert Darwin was instead a stolid seeker of prosperity, which he achieved in the style that we typically attribute to Victorians, fifty years later. To please his ever-demanding father, he studied medicine at Leiden before formally taking his degree at Edinburgh in 1786. But medicine had not been his first choice. "He for some time detested the profession, and declared that if he had been sure of gaining £100 a year in any other way he would never have practiced as a doctor."[15] Once again Erasmus Darwin had imposed his will; with the death of his older brother Charles and the continuing disappointments of Erasmus junior, Robert became his father's hope for continuing the family tradition of achievement. Predictably, Robert Darwin, too, became depressed under the relentless parental pressure, and for his whole life suffered from migraine headaches.

Despite a lifelong resentment of his father, Robert Waring Darwin inherited many of his characteristics. Like his father, he grew very portly (to be polite). In public he had the same bonhomie, but within the family his disposition was more authoritarian than sunny. He seems very much like his father: a very unsympathetic man, bullying both his family and his patients, and excessively frugal in everything to do with money.

There seems to have been a genetic disposition there. Charles Darwin observed that his grandfather "had acted towards [Robert Darwin] in his youth rather harshly and imperiously, and not always justly; and though in after years he felt the greatest interest in his son's success, and frequently wrote to him with affection, in my opinion the early impression on my father's mind was never quite obliterated." And Erasmus Darwin's own father had been just the same: he was "very tender to his children, but still kept them at an awful kind of distance."[16]

A demanding man, exacting even, Robert Darwin expected that when he was in the room, no one else would speak. His success as a doctor is perhaps surprising given the fact that he had a pronounced aversion to blood. One wonders what he did when someone brought an accident victim to his door. Everyone attributed his success to a quite psychological approach to healing and illness and his "bedside manner." But this was where an overbearing manner seems, today, often to have verged into bullying—evidently something that he had learned from his father. He had, even for the age, a distinctly patronizing manner with women and considered that most of the ills they brought to him were the result of hysterical malingering. In one anecdote, a "Lady" was brought to him suffering a mental decline, in part from feeling inadequate compared to the memory of the husband's first wife. Erasmus Darwin frankly told her that she was indeed less attractive than her predecessor and generally berated her. Then he told her husband to tell her that if she did not behave, Dr. Darwin would have to call again. (Presumably, although it was never so stated, "behaving" meant returning to conjugal relations.) Unsurprisingly, that was enough to effect a "cure"—to the extent, at least, that the poor woman had to buckle down and live with her sorrows.

Interestingly, given his own sensitivity, Darwin wrote of Robert Darwin, "My father was very sensitive so that many small events annoyed or pained him."[17]

Robert Waring Darwin married his childhood sweetheart, Susannah (or Susanna) Wedgwood, by all accounts a very intelligent, vivacious woman then (in 1796) aged thirty-one. The reason that they did not marry earlier is unclear. Perhaps it had to do with her nursing her frail, elderly father; perhaps it was related to her own constant ill health. She was the daughter of Erasmus Darwin's old friend Josiah Wedgwood, founder of the pottery in Burslem, near Stoke-on-Trent (later moved to the famous Etruria works). Their marriage thus further linked two major Midlands families of the new industrial age.

Robert Darwin quickly fathered six children in twelve years. Charles Robert Darwin (known in the family as "Bobby"), the future scientist and father of modern evolutionary theory, was the fifth. Catherine, the last, was born when Susannah was forty-five. Susannah was evidently a model wife and mother: intelligent, loving, and independent when she needed to be. She had been brought up in the Unitarianism of her father and her father-in-law and remained loyal, whereas Robert Darwin had already started a slide to the conventional respectability of the Church of England. (Charles Darwin was christened at Shrewsbury's parish church of St. Chad.) She took the children to the Unitarian chapel every Sunday (although after her death, the family attended the parish church exclusively).

In 1800, the growing family moved into an impressive new house, the Mount, with extensive land overlooking the river Severn. There, Robert Darwin is said to have taken "almost as much interest in botany and zoology as his father" and gave much of his energy to creating the grounds, hothouse, and gardens.[18] The environment that Robert and Susannah Darwin created there for their family was secure and rich without being ostentatious. There were plenty of servants, plus ponies, dogs, and a stream to fish in. Their two sons quickly developed a fondness for natural history, collecting, as small boys often do, shells, insects, and so on—but perhaps with a greater purpose and education than many.

As her health permitted, Susannah Wedgwood, a "gentle, sympathizing" woman, entertained her husband's friends and helped with his correspondence. There can be no doubt that Darwin was as influenced in his taste for natural history by his mother as by his father and the grandfather he never met. "She entered zealously into all her husband's pursuits . . . and their gardens and grounds became noted for the choicest shrubs and flowers. They petted and reared birds, and the beauty, variety and tameness of 'the Mount Pigeons' were well known in the town and far beyond."[19]

But Susannah (Sukey to her husband) never enjoyed good health. Throughout her life she suffered from intestinal problems and blinding headaches; as a child she had a bout with rheumatic fever, which probably left her with a weakened heart. Her own mother, Sarah Wedgwood (cousin to her husband, Josiah), had been seriously depressive, and Susannah became worn-out and depressed herself, and often ill and semi-invalid. The children were often ill, too: "My whole time has been taken up . . . in nursing sick children and I have still four very poorly—most violent colds, and attended with considerable fever—We are in daily fear too of the scarlet fever, it is become so prevalent."[20] Often, when Robert Darwin might have helped her most, he was away taking care of other patients.

On the surface, the young Darwin seems to have had an idyllic childhood. In addition to his siblings, there was a second family twenty miles away at Maer Hall, in the form of Uncle Jos and Aunt Bessy (Elizabeth) Wedgwood and eight cousins. These Wedgwoods were solid, foursquare propriety itself (although, by 1838, somewhat in debt to Robert Darwin, a shrewd loaner of capital).

In this large extended family of eight girls and six boys, there was much ambition, but so far little of it rested on Charles Darwin's youthful shoulders. Then calamity struck. In July 1817, after a period of many months of acute abdominal pains, Susannah Darwin died. She was fifty-two years old. Her children ranged in age from Catherine (seven) to Marianne (nineteen, and already engaged to be married); Darwin was eight.

THREE

Childhood

In many respects Darwin had a happy boyhood. Robert Darwin seems to have coped with his motherless brood by leaving the younger ones essentially to be brought up by their sisters and the family nursemaid, Nancy. Darwin reveled in all the cosseting but attached himself particularly fervently to Erasmus. He was swept along in an extended family of uncles and cousins, all achievers, mostly socially adept, personally charming, skilled in business, and expecting to succeed in everything they tried. (Erasmus, however, although charming and popular, already appeared to be indefinably weak and subject to the family tradition of depression.) There was never a time when money was a problem; it was a life of ease and servants, without the trappings of the minor aristocracy but with all the advantages of a good position in society. In other words, it was the very stuff of the novels of Miss Austen that were just then becoming great successes—and especially with regard to the need to find husbands for those Darwin and Wedgwood daughters.

But then a taste of Charles Dickens emerges in the story. Until now, Darwin had been taught at home by his sister Caroline. When his mother died, Darwin went to "Mr. Cases school" in town. "I remember how very much I was afraid of meeting the dogs in Barker St & how at school I could not get up my courage to fight.—I was very timid by nature. . . . I believe shortly after this or before I was very fond of gardening." He recalled that he was "much slower in learning than my sister Catherine."[1] The following year, Darwin was sent off to boarding

school—Shrewsbury School—whose headmaster was Dr. Samuel Butler, already famous as both an educator and a writer. Darwin's enroll-
ment there had been long planned for this, his ninth year, and Erasmus
was already there ahead of him.

Now, all of a sudden, there was no longer the security of living at
home. The only saving grace was Erasmus. Erasmus broke every path
for Charles Darwin at Shrewsbury School. The Dickensian horrors of
the school were real enough; there were rather more pupils than could
fit in comfortably. But it was all the worse for a nine-year-old whose
mother had just died. Erasmus survived better, but he was in any case
far more sociable. Erasmus's friends became Charles's friends—or,
rather, he trailed around after them.

Darwin's later assessment of the school was typically dismissive,
"Nothing could have been worse for the development of my mind than
Dr. Butler's school, as it was strictly classical, nothing else being taught
except a little ancient geography and history. . . . The only pleasure I
ever received from such studies, was from some of the odes of Horace,
which I admired greatly."[2] More to the boys' taste were extracurricular
studies: they were very much taken with a book on travel and exploration called *Wonders of the World,* and it was at about this time also that
their father borrowed from a local library the first English publication
of the journals of Lewis and Clark in their "Corps of Discovery" trek to
the American West Coast.[3]

Coming to the end of his time at Shrewsbury, Erasmus went to
Cambridge University to start training to follow the family tradition by
becoming a doctor. So, when the new autumn term came round again
after the usual long, happy family summer, Charles Darwin had to face
it essentially alone. He often ran all the way home from Shrewsbury
School to the Mount between dinner and dormitory bedtime. Nonetheless, despite hating the school, he made several lifelong friends there,
noting in his *Autobiography* with his characteristic clarity—and lack of
charity—that "not one of them ever became in the least distinguished."[4]
This was quite unfair. For example, at Cambridge he would come to
reestablish a close friendship with Charles Whitley, who later was reader
in natural philosophy at Durham University. There was an echo here of

the way in which he would later, and inaccurately, dismiss his teacher Robert Grant at Edinburgh as not having accomplished anything when he went to be professor at London. This inconsistency and lack of charity cannot simply be excused as forgetfulness. He often appeared arrogant and particularly disliked any rudeness or vulgarity in others and would then act with a coldness that was quite different from his otherwise mild disposition. William Leighton, a schoolmate at Shrewsbury, recalled that when they met again at Cambridge, Darwin's manner was "still reserved and proud."[5]

The horrors of school were probably the greater because of the contrasting pleasures of family in the vacations, when he could indulge in natural history to his heart's content. "I must have observed insects with some little care, for when ten years old [1819] I went for three weeks to Plas Edwards on the sea coast in Wales, I was very much interested and surprised at seeing a large black and scarlet Hemipterous insect, many moths (Zygaena) and a Cicindela, which are not found in Shropshire. I almost made up my mind to begin collecting all the insects which I could find dead, for on consulting my sister, I concluded that it was not right to kill insects for the sake of making a collection. From reading White's *Selborne* I took much pleasure in watching the habits of birds, and even made notes on the subject. In my simplicity I remember wondering why every gentleman did not become an ornithologist."[6]

Added to the delights of the garden and farm, between the ages of about twelve and sixteen, one of Darwin's passions was chemistry. This again was something in which Erasmus took the lead. They commandeered a wooden shed in the garden as a laboratory. To Dr. Darwin's displeasure, considerable expense was required to fit it up. Chemistry remained one of Darwin's main holiday occupations with Erasmus, even when the latter had gone to Cambridge. Darwin's schoolmates gave him the nickname "Gas." This kind of interest in chemistry was then far more than the boy's hobby that it became a hundred years later. Instead, it was a gentleman's pastime, a place where an exact science could be encountered firsthand and in a more direct, analytical, and theoretical way than in natural history. In the great age of industry and manufacturing, chemistry was (to use a much overworked term) at the cutting edge. It could be pursued at home with modest equipment, and a whole range of publications sprang up to guide the student through its secrets.

It does not take much imagination to conjure up the two Darwin boys passing many happy hours among strange and often offensive smells. But Darwin always loved the outdoors, too, and not even chemistry could keep him from his passion for collecting insects.

In trying to find adjectives to describe the young Darwin at any age from his mother's death in 1817 to his departure for university in 1825, "self-absorbed" most often comes to mind, followed by "immature." For all that he was often the rosy-cheeked outdoorsman, he was bookish, stammered badly, and could be painfully shy. He was meticulous and even obsessive about details, making lists of everything.

Darwin's introspective nature showed in his long, solitary walks, referred to earlier. He kept up the habit of walking all his life, first at Edinburgh and Cambridge with friends (and his brother Erasmus), later in total isolation on the famous "Sand Walk" that he had constructed at Down House—a place where he could think. For a man, solitary walking can be considered creative and restorative, but for a small boy, it was more than unusual.

To have come this far in trying to reach behind the public picture of Darwin to understand his childhood is already to strain against the defenses that Darwin erected to protect himself. He took great pains in later life to guard against anyone, even his family, peering too closely into his early days. He made two attempts (usually published together) at writing an autobiography, but they tell us little about his family life. The first was written in 1838, not long after the *Beagle* voyage, when he was perhaps already beginning to feel the first hints of a future fame and the first yearning for a family of his own. It is a brief document of a few fragments of recollection of his first eleven years.

Darwin revealed very little of himself in the fuller *Autobiography* that he began to write in 1876 (at age sixty-seven). He was writing for his family, at least at first, and perhaps he had forgotten a lot. Possibly it was because he was writing for his children and grandchildren that he seems to have gone out of his way to tell anecdotes about what a naughty little boy he had been, telling lies, bullying his sisters. It is easy for the arm-chair analyst to see these episodes as a calling for attention: a young boy lost in the world trying to make some kind of mark, to gain respect and love in that motherless, and soon brotherless, world.

"As a little boy I was much given to inventing deliberate falsehoods, and this was always done for the sake of causing excitement. For in-stance, I once gathered much valuable fruit from my Father's trees and hid them in the shrubbery, and then ran in breathless haste to spread the news that I had discovered a hoard of stolen fruit."[7] On another oc-casion he stole fruit from someone else's orchard in order to impress boys from the town with how fast he could run.

He described another "little event . . . curious as showing that ap-parently I was interested at this early age in the variability of plants! I told another little boy (I believe it was Leighton who afterwards became a well-known lichenologist and botanist) that I could produce variously coloured Polyanthus and Primroses by watering them with certain coloured fluids, which of course was a monstrous fable, and had never been tried by me."[8]

Was he, in his old age, still a little ashamed of this behavior? Or is it possible that in his secure, famous old age he still felt keenly that need to be recognized as a very human figure? All the stories he recounted in the *Autobiography* are oddly self-centered. When he boasted about stealing apples, this story really was a boast about how athletic he was—a theme repeated in his account of his after-dinner runs to home and back from school. All the while he mentioned nothing in the remotest way intellectual or educational except to disparage it, especially any-thing that involved learning by rote.

Surprisingly, Darwin could recall very little about his mother.

My mother died in July 1817, when I was a little over eight years old, and it is odd that I can remember hardly anything about her except her death-bed, her black velvet gown, and her curiously constructed work table. . . . All my recollections seem to be connected most with self.—now Catherine seems to recollect scenes, where others were chief actors.—When my mother died, I was 8 & 1/2 old. & she one year less, yet she remember all particular & events of each day, whilst I can scarcely recollect anything, except being sent for—memory of going into her room, my Father meeting us crying afterwards. . . . I recollect my mother's gown & scarcely anything of her appearance. Except one or two walks with her I have no distinct remembrance of any conversation, & those only of very trivial nature.—I remembered saying "if she did ask me to do something, which I said she had, it was solely for my good."[9]

The Freudian revolution and a century of psychology embolden us to look at Darwin's childhood, and the death of his mother, for signs of the origins of the oddities and inadequacies as well as the positive factors of his adult life.[10] The main reason for a search for an underlying psychology is obvious enough: throughout his life, not only was Darwin an odd, intense person, he was subject to all kinds of illnesses, at least some of which have the appearance of being psychosomatically induced. During the *Beagle* voyage he was a dashing, daring explorer, happy and confident. In middle age, while achieving scientific fame, he was characterized by querulousness, reclusiveness, and endless illnesses. All can all be connected with periods of high stress.

Psychologists and psychiatrists are left free to argue whether Darwin's apparent absence of memory of his mother is an important clue to these conditions and represents a deliberate suppression—an attempt to avoid thoughts of someone who betrayed him by leaving. Perhaps, but to what extent and to what result? Darwin himself said that his lack of memories of his mother was due to the fact that his sisters would never speak of her, changing the subject when her name came up. Being older, they had had a fuller experience of the trauma of their mother's illness and a deeper sense of personal grief at her death.

Early bereavement was not unusual in families of that time; the younger children in families often did not have a mother into their adulthood. Under an unbroken series of pregnancies, women's health broke down far earlier than that of their husbands. The children of the first wife were very often brought up by a second. Darwin's mother died when he was eight; Robert Waring Darwin's mother died when he was four. What seems different in the case of Darwin and his mother is that he was unable to grieve, to mourn the loss, to remember the illness, or to rejoice in the good times. He suppressed almost all memory of his mother. Bowlby, in his biography of Darwin, makes the point that the fact of Darwin taking long, solitary walks without any awareness of what he was thinking (a "fugue" state), suggests "a state that is known to occur in persons who have failed to recover from a bereavement."[11]

The argument has been made that in households like that of the Darwins, children did not see much of their parents anyway, but were brought up by nannies, in nurseries away from the main flow of the household. As Janet Browne says, "Susanna Darwin may never have been very prominent in the younger children's day-to-day existence." If she had been a remote sort of parent, the loss of the mother might not have been "as disruptive or devastating as might be supposed."[12] However, even though she was ill more often than not, Susannah Darwin's own accounts of nursing her sick children showed that she was not an inattentive or distant mother who left the work to others. If that had been the case, we would have to explain why Darwin's younger sister, Catherine, had vivid memories of her mother's final illness and death, while Darwin did not.

Unless he was already deemed unusually sensitive, the family would not have protected Darwin from the traumas of his mother's last days while at the same time exposing his younger sister fully to them. If we argue that Catherine might have been remembering only what had been told her afterward (but had not directly experienced), why was Darwin not also included in those family reminiscences (which he says were, in any case, forbidden)? The family, as a whole, seems to have suppressed its grief, or at least to have held it very privately. Darwin may even not have known *how* to grieve for her.

All we have are questions. Was it the loss of his mother and the subsequent—perhaps suffocating, perhaps liberating—care of his sisters that made Darwin always unnaturally dependent on others? Similarly, did the remoteness of his father cause him to be particularly dependent on his brother Erasmus? Perhaps it was this upside-down world of sisters and cousins that kept him unusually innocent of relationships with females his own age. More immediately, was it his mother's illness and death that made him a lifelong hypochondriac?

At some point after his mother's death, Darwin began to suffer from attacks of an eczema-like dermatitis, sometimes erupting on the hands and sometimes on the face, especially the lips. Under stressful conditions Darwin would also suffer strong headaches. Anxiety also produced an "upset stomach," most often at breakfast. Robert Waring Darwin had suffered in the same way. If, as an intelligent and observant boy, Darwin came to associate his own ailments with the illnesses that eventually killed his mother, that would have exacerbated any problem, so that any perfectly normal illness would become magnified in his mind into something potentially lethal.

When his face and lips erupted, he would avoid all company, canceling appointments even with friends, and hiding his misery at home. These eruptions were always "accompanied by feelings of shame and depression."[13] For example, in 1829, he planned a trip to Edinburgh to meet old friends and, as we shall see, to revisit a scene that had caused him much grief. But he stayed at home because "my lips have lately taken to be bad, which will prevent my going."[14]

Among early commentators of Darwin's youthful state of mind and health was Edward Kempf, fully forty-three pages of whose *Psychopathology* (1931) were devoted to a case study of Darwin. He related many features of Darwin's character and illnesses to a morbid interest in sex: "Darwin's interest in the sexual function is to be seen in the titles of his books, such as 'The Descent of Man, and Selection in Relation to Sex,' [and] 'The Effects of Sex and Self-Fertilization in the Vegetable Kingdom.'"[15] Given the number of his children, sex was obviously something that Darwin did not exactly suppress, but Kempf saw him as a repressed homosexual.

Kempf also saw pathology in the fact that Darwin's friend Leighton reported that Darwin's mother taught him "how by looking at the inside of the blossom the name of the plant could be discovered."[16] A little reading would have revealed to the good Dr. Kempf that Susannah Darwin and her husband were both keen botanists and that she had been reading about the work of none other than the famous Linnaeus. Linnaeus founded his system of classification of plants on the structure of the flowering parts. Linnaeus's emphasis on the fact that plants had sex lives was distressing to many eighteenth-century readers, just as it was fascinating to others. Erasmus Darwin had notoriously featured it in his 1791 poem "The Botanical Garden." Evidently, Susannah Darwin had been showing a boy with an inquiring mind that there was a system and logic to the structure of flowers. She had been explaining to him how the arrangement of a flower's parts—the stamens and pistils, anthers and ovaries—revealed to what group of plants it belonged—a rose or an anemone, a cowslip or cow parsley.

Oddly, Darwin did not reveal the extent of both parents' interest in biology or mention the minor menageries that they evidently maintained. For someone whose later ideas depended in no small part on the experimental breeding of pigeons, the omission of any mention of his parent's pigeons is strange indeed.

The *Autobiography* and a few family letters give us a tantalizing picture of the very young Darwin. On the one hand, we have pre-Victorian prosperity and comfort, success and a protective and indulgent circle of sisters and girl cousins. On the other, we have a motherless family, an ambitious, bullying father, two famous grandfathers, a weak brother, one uncle a suicide, a miserable school. At the center of it all was Darwin, still largely unnaturally dependent on his older brother Erasmus. Younger boys often adore their older brothers but, in the history of this particular family, Erasmus provided an unusually strong additional role as mentor as well as friend, filling the gaps between the distant father on the male side and the dead mother and protective sisters on the female side.

It might not have been a particularly unusual childhood except in one respect: Darwin was exceptionally intelligent and in equal measure lacked discipline where schooling was concerned. In all this, the budding scientist is not hard to find; it is shown in the meticulous lists and avid, equally orderly, collecting. But that could have been the childhood of a dozen men whose later life turned to banking, racing horses, embezzlement, or the church. If it is not asking too much of a fifteen-year-old, where was the intellectual? Grasping at straws, perhaps, we can note the avid reading, the interest in Euclid, the passion for teaching himself *and* the rejection of the classroom, and the impatience with, and intolerance of, all that was second-rate. And, always, the egotistical self-absorption and that ferocious intelligence.

Robert Darwin had long since decided that both of his sons would become physicians. Once Erasmus had gone off to university, Dr. Darwin got Darwin started, too. "I began attending some of the poor people, chiefly children and women in Shrewsbury. I wrote down as full an account as I could of the cases with all the symptoms, and read them aloud to my father, who suggested further enquiries, and advised me what medicines to give, which I made up myself. At one time I had at least a dozen patients, and I felt a keen interest in the work. My father . . . declared that I should make a successful physician,—meaning by this, one who got many patients."[17]

The 1824-25 school year had ended with Darwin still performing poorly. The problem was certainly not stupidity. Something then had to be done; if Dr. Butler could not bring out the best of an obviously intelligent young man, Charles must go off on his own. As he had a poor training in classics, Cambridge was not a good choice, but for medicine he could go to the University of Edinburgh, like his brother, father, grandfather, and namesake uncle. At sixteen, Darwin would be slightly younger than most entering students. With his keen eye for matters of temperament, Dr. Darwin realized that Darwin would not do well at Edinburgh if he went off alone. The timing was also good for Erasmus to take a year off from Cambridge to study at Edinburgh. He could act as Darwin's alter ego once again.

Edinburgh

The two brothers arrived in Edinburgh a week early, so as to prepare for the year by "reading like horses," as Erasmus rather oddly put it.[1] Staying first at the Star Hotel on Princess Street, they shopped around for rooms in the many tenement buildings near the university.

We got into our lodgings yesterday evening, which are very comfortable and near the College. Our landlady, by name Mrs. MacKay, is a nice clean old body—exceedingly civil and attentive. She lives at 11 Lothian Street [the house was later demolished to make room for part of the National Museum of Scotland] & only four flights of steps from the ground floor which is very moderate to some other lodgings we were nearly taking—The terms are £1, 6/ [presumably per week; around £85 or $170 in today's terms]—for two very nice and light bedrooms and a sitting room; by the way, light bedrooms are very scarce in Edinburgh, since most of them are little holes in which there is neither air nor light. We called on Dr Hawley [a friend of Robert Darwin] the first morning, whom I think we never should have found had it not been a good natured Dr. of Divinity who took us into his Library & showed us a map, & and gave us directions how to find him: indeed, all the Scotchmen are so civil and attentive, that it is enough to make an Englishman ashamed of himself.[2]

They had formally matriculated the day before: "We pay 10s. & write our names in a book, & the ceremony is finished; but the Library is not free to us till we get a ticket from a Professor."[3] In fact, use of the library required a £1 deposit for a borrower's ticket. All students were also required to inscribe their names in the "Album of the University" during the first week of every month in term. They both meant to be serious about their studies. Erasmus, five years older, was wise in an adult world of which sixteen-year-old Darwin had no inkling. He intended to take care of his younger brother, although Darwin was no longer a boy but a young man who would soon be delving into the unattractive aspects of gross anatomy, the varied fascinations of the hospital wards, and instruction in gynecology and obstetrics.

Edinburgh must have been something of a shock to these tall, serious young Englishmen, almost inseparable and sharing interests and studies to an extent few siblings achieve. In 1825, Edinburgh, a large, cold, smoky city, was, as it is now, one of the great cities of Europe with one of the greatest universities, a product of the Scottish Enlightenment.[4]

The university was particularly known for its excellence in the sciences and medicine although, if one looked closely, some of the greatness was wearing a little threadbare—a bastion of freethinking was trapped in an old-fashioned system of patronage that was slowly killing it. By 1825, the structures that had served the university well in the previous century were embattled and its professors under pressure. Much of the medical curriculum, once the pride of Europe, was antiquated. Unlike at Oxford and Cambridge, with their tutorial system, teaching at Edinburgh was focused primarily on courses of lectures. This was not to appeal to Darwin. He said in the *Autobiography*: "To my mind there are no advantages and many disadvantages in lectures compared with reading."[5]

A great deal depended on the quality of the professors, but professorships at Edinburgh had become family sinecures, and salaries were literally the proceeds of tickets sold. As a ticket for one course cost 4 guineas, or nearly four times more than a week's rent for the Darwin brothers, those professors who could attract a large audience could become extremely wealthy. As the quality of teaching of medicine declined in the university after around 1815, in response, private medical schools had grown up in the town, offering rival courses—better taught, in many cases—but university examinations still had to be passed.

Edinburgh was an exciting place, but possibly a little baffling to a sixteen-year-old who may have been there only because he was not doing well at school. Darwin had never experienced a really big city, although he had once taken a short trip to Liverpool. Instead of the Georgian red-brick warmth of the English Midlands and the manicured green fields of his uncle's estates, here, severe gray stone predominated in grand buildings and slums alike. The city, built on a series of hills whose geology Darwin would soon come to know well, had a chilly grandeur and the country around a harsh romance, very different from the soft landscapes of western England.

Never before had Darwin been thrust into the hurly-burly of intellectual debate, nor had he, despite the presence of Erasmus, had to depend quite so much on his own initiative. Instead of the slow, if painful, drone of school life and the rigid discipline of the masters, here he was with grown-ups and expected to make his own pace academically.

The academic curriculum at Edinburgh was arranged in four schools: Literature and Philosophy, Theology, Law, and Medicine.[6] Both young men bought tickets for Alexander Monro's course Anatomy, Physiology, and Pathology, and Darwin signed up for the other foundation of the medical curriculum—*Materia Medica,* the study of what today would be called medicinal treatments or pharmaceuticals. That side of the healing arts in Darwin's day was very little advanced from the knowledge of healing plants that was contained in the herbals of the seventeenth century. Materia Medica was taught by Dr. Andrew Duncan, a man right at the end of his career. Darwin also signed up for Clinical Lectures by Drs. Robert Graham and William Alison. Finally, both brothers enrolled in their favorite subject, chemistry (actually Chemistry and Pharmacy)—in a course taught by Professor Thomas Charles Hope, a flamboyant showman who ranged far and wide in his lectures, from electricity to geology. They also obtained "Perpetual Hospital Tickets" admitting them to the Old Royal Infirmary.

Not long after term started, Darwin reported to his sister Caroline his growing frustration with the quality of teaching at Edinburgh.

> Your very entertaining letter . . . was a great relief after hearing a long stupid lecture from Duncan on Materia Medica—But as you know nothing of either the Lectures or Lecturers, I will give you a short account of them.—Dr. Duncan is so very learned that his wisdom has left no room for his sense, & he lectures, as I have already said, on the Materia Medica, which cannot be translated into any word expressive enough of its stupidity. . . . Dr Hope begins at ten o'clock, & I like both him and his lectures *very* much. (After which Erasmus goes to Mr. Lizards on Anatomy, who is a charming lecturer) At 12, Hospital, after which I attend Munroe on Anatomy—I dislike him & his Lectures so much that I cannot speak with decency about them. He is so dirty in person & actions.—Thrice a week we have what is called Clinical Lectures, which means lectures on the sick people in the Hospitals—these I like *very* much.[7]

Despite his expressed distaste for lectures in general and the medical courses in particular, Darwin was a diligent student. Some of his notes from the lectures by Monro and Hope survive, showing that they were written out with care.[8] He bought a copy of the eleventh edition of Duncan's textbook *The Edinburgh New Dispensary* (1826). It was Darwin's bad luck that he had come to Edinburgh when his three main teachers were in their declining years. Andrew Duncan was only a pale imitation of his father, who had been chairman of the Institutes of Medicine at Edinburgh University from 1790 to 1821, physician to the king in Scotland, and generally a major figure in Scottish medicine. (He had been the admiring teacher of Darwin's uncle Charles, who was buried in the Duncan family plot.) Neither Alexander Monro 3rd nor his father, both of whom taught anatomy at Edinburgh, had any training in the subject. In a stubborn holdover from the days when the "surgeon" was the local barber (a man who at least knew the value of a clean, sharp blade), anatomy and surgery were added only very late to the medical curriculum, being thought not to be academic disciplines with a scientific basis.

Thomas Charles Hope was an entirely different kind of teacher. Hope's father had been professor of botany at Edinburgh. Hope started out as a brilliant research chemist, discovering the element strontium and being the first to observe that water expands as it freezes, attaining a maximum density at around 4 degrees Celsius (explaining, therefore, why the surface of a pond freezes first, and why icebergs float). But he gave up research in order to teach and was widely considered the most brilliant teacher of science in Britain. He wanted to make science more "accessible" (as today's term has it) and spent a remarkable effort on writing clear lectures and assembling apparatus to stage splendidly elaborate demonstrations of chemical processes and principles. (Oddly, he never gave any practical instruction to students.) He was a showman, a magician almost, but a rather deliberate one, much criticized for his pompous delivery. But, given the fact that many students came from Europe to hear him, his overly precise diction can perhaps be excused.

Within parts of the university, of course, Hope was denigrated—he was accused of being too much of a showman and of deliberately seeking out controversy to boost his ticket sales. Hope was a phenomenon, regularly attracting 500 students to his lectures; in 1829, the number was 575 (those ticket sales would yield around $300,000 in today's money). In 1827, he gave a special summer course of lectures for women and donated the receipts, worth £800, to the university to fund a prize essay competition (on chemistry, of course). Darwin and his brother evidently cultivated Hope's acquaintance and were once invited to dinner.

As Darwin's letter to his sister Caroline shows, while Monro in Anatomy was abominable, Clinical Lectures, taught in the hospitals with real patients—as in modern clinical rounds—intrigued Charles Darwin more. Here Darwin could see where all the interminable lecturing might be leading. This was the sort of work his father did and with which he had assisted at Shrewsbury. Anatomy, with its demonstrations of dissections, appealed to him far less. Students were not required to enroll for practical anatomy classes, which entailed an extra charge.

This was the Edinburgh of the Burke and Hare scandals, and cadavers for dissection were at a premium. The result was that he neglected it badly and later rued the fact that he never learned proper dissection technique. The best that lecturers like Monro would offer students was an ancient, grizzly object "fished up from the bottom of a tub of spirits." It was said of Monro's father that he would "demonstrate those delicate nerves which are to be avoided or divided in our operations . . . once at the distance of one hundred feet—nerves and arteries which the surgeon has to dissect at the peril of the patient's life."[9] There is no satisfactory explanation, however, of the fact that Darwin did not seek out other teachers of anatomy at Edinburgh. To have good firsthand experience of dissecting a human body he could have gone "over the road" to one of the private medical schools that had sprung up to make a good business in filling in the gaps in the university's teaching. Some of the best young scholars were teaching in these establishments, barred as they were from finding positions within the university's sinecure-dominated appointments system. But Darwin had no interest at all in becoming a surgeon.

If all this gritty—if not gruesome, sometimes positively medieval—medicine, and the gray granite city of Edinburgh itself were not depressing enough, even worse was to come. The operating room, in the days before anesthetics, proved a particular obstacle. Charles Darwin's aversion to surgery would seem more than reasonable to any twenty-first-century observer used to the principle of sterilizing, meticulously prepared doctors and nurses, superb instrumentation and, above all, anesthesia. Students were required to observe operations. To make matters worse, one of those that Darwin saw involved a child. "I . . . attended on two occasions the operating theatre in the hospital at Edinburgh, and saw two very bad operations, one on a child, but I rushed away before they were completed. Nor did I ever attend again, for hardly any inducement would have been strong enough to make me do so; this being long before the blessed days of chloroform. The two cases fairly haunted me for many a long year."[10]

Many authors have concluded that it was the sight of these operations and above all the sounds—the screams of the patients, the sawing of bones—that later caused Darwin to give up any thought of a career as a doctor. In fact, he observed those operations in the spring of 1826 but did not quit for another whole year. The fact is that one could be a doctor without performing or observing operations and one could, as his father already had shown, be a doctor despite feeling ill at the sight of blood. In those days, particularly, opening the patient was the last resort: there were few skills or techniques to do it. Amputations were common enough but could be accomplished by specialists or, notoriously, in the army and navy. A country doctor like Robert Waring Darwin could make a wonderful career without thinking of opening patients up.

Once they had paid their pounds, Erasmus and Charles Darwin became the university library's most extensive borrowers for the year 1825–26: Erasmus took out no fewer than forty-one volumes for a fortnight's loan each, and Charles Darwin took twenty-one. In addition, Erasmus had brought with him a considerable library of his own books.

Among the titles that Darwin is known to have borrowed were several that showed his attention to medical studies. These included John Mason Good's *The Study of Medicine* (1822) and Christopher Robert Pemberton's *A Practical Treatise on Various Diseases of the Abdominal Viscera* (1806). There was at least one general scientific work (Thomas Young's *A Course of Lectures on Natural Philosophy and the Mechanical Arts* [1807]) and also a number of books in Darwin's favorite subject of natural history. Among these were John Fleming's *The Philosophy of Zoology* (1822), William Wood's *Illustrations of the Linnean Genera of Insects* (1821), Robert Kerr's *The Animal Kingdom* (1792), and Samuel Brookes's *Introduction to Conchology* (1815). He also signed out Newton's *Optics* (1704) and Boswell's *Life of Johnson* (1791).

This list shows how serious a reader Darwin was and how broad was the range of his scholarly interests. Reading was a passion, and he guiltily admitted to his sister that he had read a great number of novels: "I have been most *shockingly idle, actually reading two novels at once.*"[11] Evidently, Darwin had found a ready source of nontechnical books, presumably a local lending library.

Perhaps one reason for their intense program of reading was that the two brothers had almost no life outside their studies. Erasmus, definitely the leader of the two, had taken it in his head to avoid a social life in Edinburgh. Despite having visited their father's friend Dr. Richard Hawley when they first arrived, they refused to accept introductions to people in Edinburgh who could have helped or simply amused them. They stuck rather grimly to their task. They studied, they read, they went to lectures and the occasional concert. For recreation on Sundays they went to various churches to compare the sermons. And they walked: when the weather allowed, they explored the Pentland Hills or the one place where Edinburgh had that great advantage over Shrewsbury or Cambridge— the nearby seashore at Portobello and Leith, where the Firth of Forth, with its wonderful mudflats and marshes, teemed with wildlife.

Darwin kept a small diary reserved for natural history notes from his Sunday ramblings with Erasmus.[12] For the first half of 1826, he made extensive notes—sometimes of fish and shells collected from the seashore, most often of birds seen or heard. The first entry was for January 18, when he noted seeing a hedge sparrow creep into a hole and wondered, "Where do most birds roost in the winter?" The next day he saw yellow and gray wagtails, noting: "Diagnosis consists in the former having black legs & in being more brilliantly coloured." Among his last Scottish observations for the spring, he noted that in the cold north, swallows had not yet arrived by April 25. Then, on April 26, on his way home on the stagecoach, he spotted "chimney swallows" and a redstart, ninety miles south of Edinburgh.

Above all, Darwin and his brother continued to read voluminously. It must have been a quiet life, but perhaps it was just what the shy Charles Darwin, with his nervous stomach and acute skin problems, needed. At the end of March, Erasmus departed, having completed his studies, leaving Darwin alone in the cold, gray city. His misery must have been communicated to his sisters, because Susan wrote to him: "For this next month devote yrself to wisdom & you will be much happier."[13] Darwin might have left, too, but he stayed on to hear Hope's lectures on electricity "and I am very glad I stayed for them."[14]

Soon they were both back in Shrewsbury, where their own chemistry laboratory now fell into disuse. Erasmus had been enlisted to assist his father: "We have been making Erasmus very useful, taking him to

Doctor all our sick poor people."[15] But apparently Darwin was not needed for this work. Instead, he spent the summer in field sports and general idleness (no doubt also bolstered by a good deal of reading). There was a walking trip in Wales. Perhaps significantly, Darwin did not mention in his *Autobiography* whether he had assisted his father in his medical practice again.

This was the summer when he learned properly to shoot. Shooting, first game and then specimens for his scientific collections, would become a major part of Darwin's life. His natural history diary records his successes in shooting partridges, rabbits, and hares.[16] Darwin recorded that he would keep his boots by his bedside during shooting season, so that he could step straight into them in the morning without wasting a moment's time. During the voyage of the *Beagle,* he tired of it and left the work largely to his servant, but in his late teenage years shooting was a passion. But it was not a skill acquired easily. His cousin Bessy Galton recorded that when Darwin first went out with her father (Samuel Galton, owner of a famous gunsmith's company in Birmingham), "the birds sat upon the tree and laughed at him."[17] So, with characteristic determination, he learned to shoot very well.

Robert Jameson

After the *Beagle* voyage, the only "professional position" that Darwin held was as unpaid secretary to the Geological Society of London. He was asked to take over the position within months of his return to England and, after demurring because of pressure of work, accepted it a year later. It was clear in 1837 that Darwin was a young geologist of some note even though, as he explained, his knowledge of English geology was deficient. It is remarkable, therefore, that in his student days at Edinburgh he had resolved never to study the subject.

One can imagine Darwin, in October 1826, dreading the return to Edinburgh and the renewed drudgery of the medical curriculum without Erasmus as companion and buffer against the outside world. He took a single room down the street from their old quarters and began his studies again, taking tickets for Dr. Home's Physic (Practice of Medicine) course and Dr. Hamilton's Theory and Practice of Midwifery, together with a six-month ticket for the Edinburgh General Lying-in Hospital.[1]

Without his main rudder, Erasmus, he sought out new friends. It is characteristic of Darwin in his Edinburgh career that he did not look for them either among those students most keen on a medical career or among the more idle sets of sportsmen or gamblers.

> My brother stayed only one year at the University, so that during the second year I was left to my own resources; and this was

an advantage, for I became well acquainted with several young men fond of natural science. One of these was Ainsworth, who afterwards published his travels in Assyria; he was a Wernerian geologist, and knew a little about many subjects. Dr. Coldstream was a very different young man, prim, formal, highly religious, and most kind-hearted; he afterwards published some good zoological articles. A third young man was Hardie, who would, I think, have made a good botanist, but died early in India. Lastly, Dr. Grant, my senior by several years, but how I became acquainted with him I cannot remember; he published some first-rate zoological papers, but after coming to London as Professor in University College, he did nothing more in science, a fact which has always been inexplicable to me.[2]

These new friends were all members of the Plinian Natural History Society; another friend, William Kay, was its secretary. Whereas in the previous year, he and Erasmus had not been joiners, right at the beginning of the new term Darwin became a member of this group, a student scientific society that met on alternate Tuesdays in term time. Usually not more than twenty or so members attended. The Plinian, like many similar groups of the age, was much set about with structure, committees, officers, and rules. Only a week after being elected to membership, Darwin became a member of the council, suggesting either that the other members were lax or that Darwin presented a particularly keen new face.

With Kay, Darwin wrote for his Plinian colleagues a partly fictional account, "Zoological Walk to Portobello." This was Darwin's first formal attempt at humor and it reads pretty much as one would expect of undergraduate writing by Englishmen away from home. Their long-anticipated expedition was plagued by distinctly Scottish foul weather; their haul of natural history observations was limited to finding a few miserable shells. "But if we failed in adding to our stock of Zoological Knowledge we had, on our return, the satisfaction of discussing in a most scientific manner an excellent dinner, not of Haggis or Scotch Collops, but of substantial Beef-steak."[3]

The fact that Darwin joined the Plinian Society is significant for a number of reasons. Perhaps the least obvious is that it represented several kinds of maturing in the young man. He was trying to overcome his shyness and meet people. He was trying to join the intellectual life of the university in a way different from most undergraduates, and this shows that his interest in science was not just that of the book learner, simply trying to pass examinations, and extended beyond the realm of medicine. He was showing his most fundamental character—as observed by Uncle Jos a few years later, he was a man of "enlarged curiosity," and that curiosity was totally removed from anything to do with medicine.[4]

Darwin also received invitations to meetings of the Royal Society of Edinburgh, including one at which Sir Walter Scott spoke, and to meetings of the Wernerian Society. Membership in the Wernerian Natural History Society, which devoted itself to discussion of the active scientific issues of the day, was open only to graduates of the university. With all this, Darwin was close to the center of scientific life in Edinburgh. In fact, in joining Edinburgh's scientific society, Darwin found himself thrown (or to have jumped) into the cold deep end of a disputatious time, surrounded by the tensions and difficulties not only of professors arguing intellectual matters but also various attacks on the university itself, leading eventually to a formal Commission of Inquiry.

It seems that Darwin originally did not sign up for the one course of lectures that, beyond all others, might have been expected to attract him. But, perhaps influenced by his new student friends and their enthusiasm for natural history, Darwin changed his mind within the first month of term and decided to take Professor Robert Jameson's course in natural history.[5] And with this, we can begin more surely to trace the evolution of the scientific and intellectual Darwin and to follow the origins of his ideas on the transmutation of species.

Robert Jameson, professor of natural history from 1805 to 1844, was the leading scientist at Edinburgh in Darwin's time and in every sense the dean of natural science in a university for which science was important, and perhaps more than important—central. His natural history museum was one of the largest in Europe. He was the author of the standard textbook of mineralogy. He had also founded the Wernerian Natural History Society as a place in Edinburgh where medical scholars could meet and exchange ideas on science. Jameson was also the editor of several English editions of the *Discours préliminaire* of Georges Cuvier's great work, *Récherches sur les ossemens fossiles de quadrupèds.* This Jameson renamed *Essay on the Theory of the Earth* and to it, in a most extraordinary fashion, he added a catalogue, "Geological Illustrations," in the form of a series of miniessays far longer in total than Cuvier's own text.[6]

Jameson was the editor of the *Edinburgh New Philosophical Journal*—one of the leading scholarly magazines of the time—and had been coeditor with David Brewster of its predecessor, the *Edinburgh Philosophical Journal.* In these journals the sciences of the day were paraded and debated in essays by the greatest men in Europe and America (not a single article by a woman, of course). Important foreign-language works were excerpted and translated for the English-speaking reader. These journals are a true time capsule of early nineteenth-century natural science. Jameson edited with a view to balance. Packed in the pages of these journals were letters from Ross in the Antarctic and Franklin and Parry in the Arctic, notes on the climate of places as far-flung as Seringapatam and as close as Leith, suggestions for improvements to steam engines, and discussions of the "laws of Electro-Magnetic Action."

To this day, one cannot open an issue of either of these old journals without finding something utterly fascinating and still topical. These journals reflect the fact that, in the first two decades of the new century, Edinburgh had become the most intellectually exciting place for science in Britain. Whereas Oxford and Cambridge had somewhat insular viewpoints and their curricula and academic staff were strictly under the control of the Church of England, Edinburgh was free in all senses. Its natural connections, therefore, were with the great universities of continental Europe.

At Edinburgh in 1826, as indeed had been the case in 1800, natural science began, and usually ended, with geology. Edinburgh was not only the home of Jameson, one of the leading mineralogists and scientific authors of the day; it had been home to the pathbreaking geologist

James Hutton (1726–97), whose theories of earth processes were revolutionizing the subject of geology. By 1800, a great deal was known about the structure of the earth, layer by layer, formation by formation, its metamorphic and sedimentary rocks and their mineral composition, and its fossils. The rocks of the earth's surface were steadily being read like a book, page by page, and the story that book told was one of deep time and of constant material change.

It was history, a natural history, on the grandest scale. Every year the earth appeared older as the scale of the processes involved in its formation and modification became more clear and as scientific thought uncoupled itself from a rigid biblical view of Creation. In the study of geology, one could search for the laws that controlled change in the earth—just as there are laws to control the movement of the planets or the structure of crystals. As geology, then, came to stand at the forefront of the advancing sciences, there were many long-standing questions to be answered. What was the original nature of the earth? What was the cause of its present, incredibly complex structure? How were mountains formed? How, and how many times, did life appear on earth?

Some things were quite obvious. One could understand, for example, that the action of rain and frost broke rocks down, and the resulting fragments of sand and clay were carried by rivers to the sea. One could see the buildup of mudflats like those on the Firth of Forth. One could see that sedimentary rocks must have been formed in the past from similar processes. Did that mean that the earth was constantly being eroded, until at some point it would all be flat and featureless? What did it mean that the apparently oldest rocks on earth were found in the high mountains? By 1800, several rival theories had been proposed, by means of which the various strands of evidence could be woven into a "theory of the earth." Edinburgh was one of the places where these theories were most energetically, if sometimes bitterly, debated.[7]

Jameson was never popular. A dour, aesthetic Scot whose approach to pedagogy was the same as his scientific style—the close ordering of knowledge— Jameson gave his students a superb review of the materials of geology and natural history and a familiarity with the best classificatory systems (whether of minerals or insects). Some students found all this cataloguing of nature fascinating. To others it was merely supremely dull. Despite being so eclectic in his interests, Jameson somehow gave the impression of being narrow-minded—probably because he held his scientific positions so vehemently.

Jameson had studied at the School of Mines at Freiberg with Jacob Gottlob Werner, one of the greatest geologists of the previous century. Werner had essentially founded the science of systematic mineralogy and had very firm ideas on the subject of the composition and early history of the earth. And this brings us to one of Darwin's first exposures to a major scientific controversy.

Wernerians subscribed to the then-popular view that the oldest rocks on earth—the "Primitive Series" (what we term the Precambrian)—had been created as a result of processes that occurred in water. As Cuvier summarized the issue: "According to [Neptunists], every thing has been successively precipitated and deposited, nearly as it exists at present; but the sea, which covered all, has gradually retired."[8] Not only did this view accord with contemporary mineralogy, the subject that Werner dominated in Europe, it was also a theory that was consistent with the biblical view of the earth as having been a formless, watery void before God created dry land on the third day of Creation.

Neptunists did not deny the phenomena of volcanism, as demonstrated in new and ancient lava flows, but they stuck to a narrow view of the formation of early rocks and minerals in water. In their view, mountains represented the very oldest rocks on earth, left standing after the rest of the surface had been eroded away by rivers and floods, to be deposited sequentially, lower down in the plains; all granites were formed in water. For a long time, "Neptunist" theories of earth history dominated European geognosy, until they died from the weight of their internal improbabilities and contradictions and were slowly buried by a range of newer "Vulcanist" theories. Those theories began with James Hutton, another Scot, whose works were widely popularized after his death by John Playfair, professor of mathematics at Edinburgh.

James Hutton was a polymath, one of the great giants of the Scottish (indeed of the entire) Enlightenment. Hutton, who studied both chemistry and medicine, ventured into farming, chemical manufacturing, and geology but was always a philosopher. He followed scholars like LaPlace, who argued that the earth had been formed as a fiery, molten mass that had slowly cooled, causing rocks to crystallize out. He proposed that the dominant forces shaping the earth still continued to be driven by heat. In Hutton's view, two processes operated in the earth. While erosion was constantly wearing down mountains and depositing silt, which built up to form new sediments, the surface of the earth was also being actively elevated into new mountains by the action of volcanoes and earthquakes driven by this inner heat of the earth. Neither of these ideas was, by itself, particularly new. Hutton's genius was to put them all together such that the sum of geological processes created a balance of offsetting erosional and uplifting forces. Mountains and plains, in this view, are formed as part of such a dynamic set of processes that he even used the metaphor that the earth was "alive."

Hutton argued that all the forces that had shaped the earth over geological time were essentially of the same nature and scale as those that we see acting today (uniformitarianism). Cuvier, on the other hand, had given his authority to the opposite view and claimed: "We shall seek in vain among the various forces which still operate on the surface of the earth, for causes competent to the production of those revolutions and catastrophes of which its external crust exhibits so many traces."[9]

Because the processes envisioned by Hutton operated on a minor scale relative to the size of the earth, this also meant that the earth was very, very old. Hutton had originally thought that, if he could measure the rates of processes by which the earth currently changed (erosion and sedimentation perhaps being the easiest), he could actually extrapolate to calculate the age of the earth. Instead, he concluded that the earth had passed through at least three full iterations since its origin; the original rocks making up the earth's surface were long gone. He ended his classic essay *Theory of the Earth* (1788) with a paragraph that has come not only to be one of the most beloved in science but also a keystone of debates over the likelihood of reconstructing a full history of the earth. "Here are three distinct successive periods of existence, and each of these is, in our measurement of time, a thing of indefinite duration. . . . in nature there is wisdom, system, and consistency. For having, in the natural history of this earth, seen a succession of worlds, we may from this conclude that there is a system in nature; in like manner, from seeing revolution of the planets, it is concluded, that there is a system by which they are intended to continue those revolutions. But if the succession of worlds is established in the system of nature, it is in vain to look for anything higher in the origin of the earth. The result, therefore, of our present enquiry is, that we find *no vestige of a beginning,—no prospect of an end.*"[10]

By 1826, most geologists were edging to the Vulcanist and uniformitarian views and were trying, where possible, to reconcile the two approaches. A particular point at issue in contemporary geology remained the origin of the rocks called basalts, and one did not have to walk very far from the university to find a classic example. Edinburgh is itself something of a living textbook of geology. The skyline is dominated by Arthur's Seat and the Salisbury Crags. In modern interpretation, Arthur's Seat is the remains of a roughly 375-million-year-old volcano; Edinburgh Castle, looking down over the city, sits on the remains of one of the cone's plugs. Both it and Salisbury Crags, standing to the east of the city center, show signs of wholesale erosion.

Salisbury Crags are the remains of a "sill"—a lateral intrusion of volcanic material into older sedimentary rock—and were formed some 25 million years after Arthur's Seat. The dense gray rock ("whin") of this sill was extensively quarried for building stone and cobbles for Edinburgh's streets. The sill contains its own, more vertical, intrusions or trap dykes of basaltlike material. The crags therefore record a number of different events occurring at different times. When he examined the quarries in Salisbury Crags, James Hutton discovered one of the important pieces of evidence for his theories. He saw that the "dykes" had themselves once been molten and had been intruded into the sill long after the sill itself had cooled. One could even see how the heat had altered the adjacent older rocks.

Neptunism and Vulcanism (or Plutonism) were theoretical constructs that, in the nineteenth century, could be tested only by mineralogy—discovery of how the constituent rocks had actually been formed. At the time Darwin reached Edinburgh, another factor in analyzing the structure of the earth was growing in strength—the fossils contained in sedimentary rocks were also becoming important in the debate over the history of the earth and the history of life on it.

It had long since been established that the earth had been populated in the past by organisms that were now extinct. The theological significance of this last fact alone was great: God had not made a perfect world or, at least, he had frequently changed his mind. Had he perhaps made mistakes? The great fact of the fossil record was that very many kinds of animals and plants had become extinct, and the likelihood that the Noachian Flood accounted for the existence of fossils was rapidly becoming remote. Even the geological credibility of the Flood was under attack.

In the study of nineteenth-century geology, while the rocks themselves were, of course, paramount, it was the fossils contained within them that most clearly showed a pattern of change—from simple to complex and, as it was thought for a long time, from warm to colder climates. Thus, in the coal measures one could find fish and reptiles but no mammals and birds. And coals were formed in tropical swamps in now-chilly places like Midlothian and Yorkshire. Going back to the Devonian, there were apparently no land plants, but plenty of fish. In older and deeper rocks still, there were no fish and the only plants were simple mosslike organisms. As for humans, no human fossils had been found at all, which seemed to confirm that humans had been created on different terms than the rest of nature.

One of the most crucial conclusions to be drawn from the emerging fossil record was that the history of the earth and life on it was one of continuous change. This contradicted the view, apparently documented in the book of Genesis, that there had been but a single point of Creation, and a single catastrophic Flood, after which everything was set in place until Judgment Day. Cuvier, among others, had tried to supplant this idea with the hypothesis that there had been a series of worldwide catastrophes (accounting for various fossil-bearing strata in which the constituents of life on earth had been replaced). But it was becoming clear that life had changed more continuously, and if through catastrophes, then these must have been millennial or even yearly in occurrence and small in scale.

It was not until the last three decades of the twentieth century that some of these major issues were settled. Mountains are built, we now know, because of the buckling of tectonic plates—a mechanism that would have fascinated both Jameson and Hutton, and perhaps horrified them both. Instead of a passive, static earth, occasionally convulsed by catastrophes, the study of geology has revealed us as the inhabitants of a roiling, churning planet driven in slow motion by the movement of semimolten rocks only a few miles at best beneath the fragile crust on which life is supported.

At Edinburgh, with Jameson traditionally cast as a Neptunist in the mold of Werner and his European precursors and Hope as a Vulcanist of the Hutton-Playfair school, the debate between the two worldviews had been long running. They debated in their classes, mercilessly assaulting each other—all of which, of course, was good for the box office, the sale of the tickets upon which their livelihood depended. Jameson nonetheless published the views of his intellectual opponents in his journals. And, right in front of them all, every day, were the Salisbury Crags, which came to be a perfect symbol of the whole issue. Were the trap dykes, as a Neptunist would believe, formed of sedimentary material filled in from above? Or were they, as the Vulcanists insisted, forced up from below in molten form?

As far as geology was concerned, Darwin could hardly have been a student at a time when natural science in Edinburgh was more interesting, either as a "philosophical" subject (the investigation of fundamental causes) or as a "historical" subject (natural history and the acquisition of new information). He attended Hope's lectures with great enthusiasm and there listened to Hope's passionate espousals of the Huttonian approach to geology. However, when it came to Jameson himself, he recorded in the *Autobiography* that "During my second year at Edinburgh I attended Jameson's lectures on Geology and Zoology, but they were incredibly dull. The sole effect they produced on me was the determination never so long as I lived to read a book on Geology or in any way to study the science." This does not jibe with the fact that he felt sure that he was "prepared for a philosophical treatment of the subject" because of his childhood interest in rocks and minerals.[11]

Both mineralogy and zoological classification had their foundations, first and foremost, in describing and classifying the minutiae of data. Perhaps the level of detail required to master mineralogy exceeded a young man's capacity to concentrate on the big issues. Even so, given his long interest in both chemistry and mineral collecting, and in natural history in general, it is surprising that Darwin could have found Jameson's course so "incredibly dull." When we remember that ten years later (five of which were spent in isolation on the *Beagle*), Darwin became a serious geologist, his harsh dismissal of Jameson is even more surprising. But it is worth noting that he did not say that he learned nothing from Jameson; he reserved to himself an interest in a "philosophical treatment" in geology. In that case, it is important to discover what exactly Jameson taught Darwin in the year 1826–27.

There is a curious footnote to be added here. In 1844, Darwin wrote to the Swiss geologist Adolf von Morlot, who had asked for help in getting some work published. Darwin's reply is a polite brush-off in which he said that Morlot "*might possibly*" have luck with the *Edinburgh New Philosophical Journal*, but "I am not acquainted with the Editor."[12] The editor was still Jameson, as Darwin would have known perfectly well. Darwin's remark is not, as one author has suggested, an indication that Darwin "never knew Jameson personally at all."[13] Among possible explanations, perhaps he did not know whether Jameson (then aged seventy) was *still* the editor; perhaps he used a "white lie" in avoiding involvement in von Morlot's problems, or perhaps it is yet another example of his claiming an intellectual distance from Jameson. Most likely, Darwin had simply forgotten. A parallel curious example of Darwin's lapses in memory occurred in 1831, when he wrote to his cousin and close friend William Darwin Fox, who wanted to visit the entomologist James Francis Stephens in London: "I utterly forget both Stephens Christian name & direction [that is, address]."[14] However, Darwin both used Stephens's publications regularly and had visited him at his home in London twice in the previous two years.

Jameson probably never did anything easily. Intense and brilliant, he was a collector—of specimens and of information—and a synthesizer, rather than a great scholar. His influence was enormous, just as his knowledge was encyclopedic. And if his influence sometimes depended on the opposition it provoked, thereby stimulating opposite views, he would not be the last to occupy such a position in Darwin's intellectual development.

Jameson's course consisted of about a hundred lectures, given five days per week for five months starting in November and ending in April. With news bulletins of discoveries in natural sciences arriving from all over the world for the journals he edited, Jameson was perfectly positioned to keep the content of his natural history course up-to-date. We do not have a precise description of the course as it was given in the year when Darwin took it, but in a later memoir Sir Alexander Grant reported that in his zoology lectures Jameson had begun with the "natural history of man" and ended with "lectures on the 'Philosophy of Zoology' of which the first was on the origin of the Species of Animals."[15]

Luckily, a record of the first half of the course, as taught in the year 1822–23, survives; these notes were made by W. D. Wilson, who wrote down in a manuscript book "a few of the general facts,—and any thing of interest not to be found in the text books of the class."[16] Two such textbooks were mentioned specifically: Charles Stewart's *Elements of Natural History* (second edition 1817) and Henry Robertson's *A General View of the Natural History of the Atmosphere and of Its Connection*

*with the Sciences of Medicine and Agriculture; Including an Essay on the
Causes of Epidemical Diseases* (1808).[17] Students in later years were di-
rected to the book of another Edinburgh colleague, Dr. John Stark.[18]
Students were expected also to buy Jameson's own *Manual of Mineral-
ogy* and to be familiar with his current edition of Cuvier's *Essay on the
Theory of the Earth* (Darwin bought it). In addition, students were re-
ferred to timely articles in Jameson's two journals and the *Edinburgh
Journal of Science.*

Jameson had also created a magnificent natural history museum at
the university and students were expected to spend afternoons there,
studying. Lecture notes and diary entries made by the naval surgeon
Robert McCormick (later to be a shipmate of Darwin on the *Beagle*),
who enrolled in Jameson's course for 1830–31, suggest that museum
study was concentrated in the second term, after Christmas.[19]

Darwin attended Jameson's lectures regularly and was diligent in
the compulsory sessions in the museum, with its synoptic displays of
insects, birds, fish, minerals, and rock types. "I became acquainted with
the curator of the museum, Mr. McGillivray (later professor at Ab-
erdeen), who afterwards published a large and excellent book on the
birds of Scotland. He had not much the appearance or manners of a
gentleman. I had much interesting natural history talk with him, and he
was very kind to me. He gave me some rare shells, for I at that time collected marine mollusca." Darwin also met a "*negro* [who] lived in Edinburgh, who had traveled with Waterton [Charles Waterton, a naturalist
who had explored widely in the West Indies and South America a decade or so earlier] and gained his livelihood by stuffing birds, which he
did excellently; he gave me lessons for payment, and I used often to sit
with him, for he was a very pleasant and intelligent man."[20]

It has been noted that the American Asa Gray, and others who took
Jameson's course in the 1830s, did find Jameson an "old dry, brown
stick." But even allowing for differences in personal taste, Wilson's
notes from the 1820s, in emphasizing the elements not mentioned in
textbooks, describe a course of lectures that (for the first term, at least)
must have been fascinating. Jameson roamed over a wide range of subjects, usually with reference to medicine but often ranging into matters

of great general interest and constantly laced his presentation with odd facts and fascinating anecdotes. Not dull at all.

One unusual, and most likely unattractive, feature of Jameson's course is that, after some introductory lectures, he launched into a curious bipartite mode of teaching. During the 1822–23 session, he lectured on physical science and zoology on alternating days. Thus on November 21, 1822, he began a series of lectures on zoological subjects by discussing the "differences between man and the lower animals . . . differences in the races of man . . . and differences among the individuals and populations of earth." Here Jameson introduced his notion of four races: European, Mongolian, Ethiopian, and American. The very next day, he began his treatment of meteorology by lecturing on atmospheric pressure and the value of the barometer, a "most useful instrument, indicative of weather, [that] has been used since the time of Boyle. It is always very low before a great storm or earthquake. . . . Barometers are now kept on every well appointed ship—use of them exemplified by reading an extract from Capt. Scoresby's *Voyage to the North Seas.*"[21]

This pattern continued when on Monday, November 26, he again lectured on the races of man and then, on the 27th, discussed climate and its effects on plant distribution (presumably referring to the work of Alexander von Humboldt that he had been publishing in the *Edinburgh Philosophical Journal*). On the 28th, he spiced up his lectures on man by mentioning the existence of "a six fingered race of men, and a six toed and it is the opinion of many that the breed . . . might be preserved." Most of this lecture was concerned with dynamics of mortality, longevity, and the oddities of sex ratios in human populations. Wilson notes a reference here to "a curious paper on this subject in the *Edinburgh Philosophical Journal.*" In fact, there were two such papers: Jameson had reported on the work of "M. Huefland of Berlin," "Remarks on the Comparative Number of Sexes at Birth" in 1822, and the next year George Harvey published "Remarks on the Increase of the Population of the United States, and Territories of North America, with Original Tables." Sex ratios were also mentioned briefly at the time in a book by Dugald Stewart (professor of moral philosophy at

Edinburgh) entitled *Elements of the Philosophy of the Human Mind* (1822). All authors noted the statistical fact that the ratio of boys to girls in the children of a single family was rarely 1:1, whereas in the population at large it always was equal. Wilson's notes state: "Excess of births from each marriage in civilized country average of 4 children. . . . In Europe, more males are borne than females in proportion of 21 to 20, but mortality of males is greater—or 27 to 26."

That population dynamics were interesting to Jameson is evidenced by the fact that he subsequently published another paper on the same subject in his new journal: "The Change Which the Laws of Mortality Have Undergone in Europe within the Last Half Century, or from 1775 to 1825, by M. Benoiston de Chateauneuf." The whole subject was, of course, illuminated by Thomas Robert Malthus's *Essay on the Principle of Population,* published in 1798 and, among others, by Benjamin Franklin's observation in 1751 that populations in North America had increased at a rate of doubling every twenty-five years, which was the rate that Malthus later assumed to be the maximum for humans. The added significance of all this is that Malthus's ideas on population were central to the later development of Darwin's theory of natural selection.

On the next day (November 29), Jameson started out on comparative zoology by laying out the structure of the vertebrate and invertebrate phyla (in his notes, Wilson refers to these as the Vertebrosa and Invertebrosa). In his first lecture on mammals, Jameson covered the apes. This was timely in that an orangutan specimen had recently been collected in Borneo. Jameson referred his students to the description by Dr. Jeffrey of Boston in the *Edinburgh Journal of Science* of the animal's transport on a ship from Batavia. The report was little short of sensational: "In external appearance, he resembled an African, with the neck somewhat shorter, and the neck projecting forward. He was three feet and a half in height. . . . While sitting at breakfast [Captain Blanchard] heard some one enter a door behind him, and found a hand placed familiarly on his shoulder; on turning around he was not a little surprised to find a hairy negro making such unceremonious acquaintance with him. George, by which name he passed, seated himself at table by direction . . . and, after partaking of coffee, &c. was dismissed.

He kept his house on ship-board clean and, at all times, in good order. . . . he would sometimes become so rough, although in good temper, as to require correction . . . on which occasions, he would lie down, crying very much with the voice of a child, as if he had been sorry for having given offence."[22]

And so the term continued, through atmospherics, meteors, mammalian characteristics, and so on. Alas, Wilson's notes end with Jameson's lecture of January 6, 1823, on the hydrology of lakes, with conjectures about the existence of "underground siphons." We do not know whether Wilson simply left the course. More likely, from this point onward, the titillating asides by Jameson decreased as he buckled down to the minutiae of animals and plants, expounded group by group, and to the equally precisely organized details of mineralogy that Darwin later found so boring.[23] However, we have a clue about the sorts of commentaries on geology that Jameson would have included to enhance his lectures. These are the "Geological Illustrations" that he had just assembled for inclusion in his fifth edition of Cuvier. Here, as in his course, he ranged widely, tackling subjects as disparate (and fascinating) as "The Subsidence of Strata," the "Deluge," "Action of the Sea upon Coasts," "An Account of the Fossil Elk of Ireland," and "Account of the Living Species of Elephant, and of the Extinct Species of Elephant or Mammoth."

A second account of Jameson's course exists in the form of the detailed lecture notes taken by Robert McCormick. In the 1830–31 session, Jameson arranged things differently, concentrating most of the subject matter into more consistent units, although still with some odd alternations of zoology and geology. In this later session, Jameson devoted most of the time to geology and mineralogy.[24] Unfortunately, however, while Wilson had made special note of the extra details with which Jameson spiced his lectures, McCormick doggedly stuck to recording the bare facts. Nonetheless, everything described by Wilson and McCormick is just the sort of eclectic science that would appeal to Darwin in later years. One cannot sympathize with Darwin if he really did find *all* of Jameson's course "dull." If he did, we would not be surprised if he had neglected mineralogy, just as he had previously neglected dissection. But

Darwin's copy of Jameson's *Manual of Mineralogy,* although the inside cover is decorated with childish doodles, is also heavily annotated in a way that shows that Darwin was paying close attention to the course and the associated work in the museum.[25] The annotations continue from the beginning to the end of the book, perfectly consistently, and they give the strong impression that Darwin used them as a substitute for taking lecture notes. The book essentially repeats Jameson's lectures on mineralogy, with which Darwin followed along, making annotations where Jameson had added information. Darwin also used the book as his guide to the specimens in Jameson's museum, making similar notes. Whether they were dull or not, Darwin evidently paid a great deal of attention to Jameson's lectures and learned his geology and mineralogy thoroughly.

. Darwin surely could not have found the raging controversy between Neptunists and Vulcanists, still being fought out at Edinburgh, dull or boring, although he might have shied away at the level of personal animosity with which it was conducted—the sort of intellectual aggression that his friends suffered at Plinian Society meetings. Each year in April, Jameson's geological lectures ended with Saturday afternoon field trips to Arthur's Seat and the Salisbury Crags.[26] Jameson took them to the very spot that Hutton had found so convincing for Vulcanism. Darwin's memory of this, in the *Autobiography,* is fairly unequivocal, at least at first glance: "I . . . heard the Professor . . . discoursing on a trap-dyke, with amygdaloid margins and the strata indurated on each side, with volcanic rocks all around us, say that it was a fissure filled with sediment from above; adding with a sneer that there were men who maintained that it had been injected from beneath in a molten condition. When I think of this lecture I do not wonder that I determined never to attend to geology."[27]

This is quite a telling passage. Jameson's supposed sneer was neatly used against Jameson: the teacher was wrong, the student right, and somehow the impression is given that the student was right all along. It is a thoroughgoing put-down, using all the right technical language: "amygdaloid margins and the strata indurated.'" (The term *amygdaloid* refers to a rock type with internal cavities within which other minerals are deposited.)

Regrettably, however, Darwin's autobiographical remarks are unlikely to be a true account of the occasion. In 1827, the year that he lectured on geology to Darwin, Jameson had retreated from his full Neptunist views, actually concluding that "secondary trap rocks, such as basalt, greenstone, trap-tuffa and amygdaloid," were volcanic in origin. He had published that conclusion in an article on rocks collected by Parry and Ross in the Arctic.[28] Furthermore, in his recent edition of Cuvier's *Essay on the Theory of the Earth,* one of Jameson's characteristic illustrative essays—"Formation of Primitive Mountains"—concerned just this subject. There he gave a concise and evenhanded review of the experiments of the German chemist Eilhard Mitscherlich (1794–1863) on the formation of ores by fusion. "The artificial production by fusion, of the minerals which compose our primitive rocks, appears, according to Mitscherlich, to place beyond doubt the theory that our primitive mountains were formerly a melted mass."[29] Jameson did not quite go so far as to endorse Mitscherlich's view completely. But he had removed from the fifth edition of his Cuvier the section that in the fourth edition had referred to the Salisbury Crags as having been once covered in water.

In the textbook that Jameson recommended to his students the following year, John Stark concluded, concerning the Neptunian and Plutonian theories, that "neither of these taken singly is sufficient to account for the present distribution and arrangement of rocks."[30] Similarly, McCormick's notes for Jameson's January 1831 lectures record that on January 10, 1831, Jameson stated: "The Sand Stone employed in Building in Edinburgh is of Neptunian or Aqueous origin. . . . The Greenstone of Salisbury Crags is of Plutonian or Gneiss origin.—as is also that of Arthur's Seat. Columnar or Prismatic Structure is chiefly confined to Rocks of Igneous origin. " On January 21, he taught that there were two sets of "Primitive Rocks," Neptunian and Plutonian in origin, and he then contrasted the two theories.

Given what Jameson had recently published, it seems impossible that his account of the rival theories, delivered to students on site at Salisbury

Crags in 1827, would have been as one-sided as Darwin remembered. Jameson might well have expressed a lingering preference for the Wernerian view, but he was past the point of certainty and sneering. In this light, Darwin's remarks might simply represent a slip of his memory. On the other hand, all the language with which Darwin discusses Jameson gives the impression of having been very carefully crafted to help create an intellectual distance from Jameson and specifically to reject the notion that Darwin's subsequent successes in geology owed any debt to him or any other teacher.

All this makes it doubly important that, as Sir Alexander Grant reported, Jameson "used to finish up with lectures on the origin of the species of Animals!"[31] That was the subject above all others in which Darwin wanted to retain all credit for his discoveries for himself.

SIX

Mentors and Models

With Erasmus gone, Darwin lacked a companion for his long country walks to the hills or the shore, and he made very few entries in his natural history diary for the autumn months of 1826. Darwin stayed on in Edinburgh for the holiday period. On December 23, he saw a gray wagtail and a water ouzel. On Christmas Day, he was out for another of his long, solitary walks: "A remarkably foggy Day, so much so that the trees condensed the vapour & caused it to fall like large drops of rain. Saw a hooded crow feeding with some rooks by the sea shore, near Leith."[1]

Of all the new friends that Darwin made in his second year at Edinburgh, perhaps none was more influential than "Dr. Grant, my senior by several years."[2] Robert Edmond Grant (1793–1874) was a member of a prominent Edinburgh family who took his degree in medicine at the university in 1814. He was a brilliant zoologist and very much in the European intellectual mainstream. He had studied in Paris and was radical in his views of both science and politics. He taught anatomy at one of the private schools in Edinburgh. Family funds allowed him to avoid practicing medicine and to devote a great deal of his time to studying the marine life of the Scottish coasts. That made him, in every way, a natural person for the young Darwin to emulate.[3]

Most people found Grant, a bachelor, a strange and difficult man, dour and morose, altogether something of a loner. With Darwin he came alive; he saw the spark of a real scholar in Darwin and opened up in the company of the young man whose enthusiasm for natural history was as ardent as his own. Grant was personally even more intense (if possible) than Jameson; his passion was the study of the lower orders of marine life, and he brought to his laboratory microscope a range of primitive sea creatures collected right there along Scotland's east coast—the simple corals, sponges, and "flustrae," the last being the British representatives of the group of organisms now called Bryozoa and found quite commonly in the intertidal waters of the Scottish coast. With their branching, apparently colonial, habit, these kinds of organisms had reminded eighteenth-century naturalists of plants as much as animals. In Darwin's time flustrae were sometimes called "zoophytes," or animal-plants; the related term Bryozoa derives similarly from the words for *moss* and *animal.*

Darwin said little in his *Autobiography* about Grant's personal life or the range of his scientific interests which, as a result in part of the time he spent in Paris, was very broad. Grant shared with the zoologist Etienne Geoffroy de St. Hilaire the view that the structure of animals could be seen as variations on a very small number of themes. Geoffroy's assistant Georges Cuvier proposed that there were four such themes: Radiata (jellyfish, corals, starfish), Mollusca, Articulata (crustaceans, insects, arachnids), and Vertebrates. Within each plan, adaptation was essentially perfect, and all the parts were fitted together so closely that species had to be invariant (or they would die). Diversity, for Cuvier, was a product of the laws of structure, all divinely created. Geoffroy eventually came to think that there might in fact be only one plan. If all animals that have ever lived are variations on this single theme, life could be seen not as a series of static species formed by the Creator, but as the result of a history of transformations and change—resulting in *adaptation* to particular life conditions. Grant supported Geoffroy's side in this debate between the two colleagues in Paris and was active in this issue during the time that Darwin was at Edinburgh working with him.[4]

In his own research, Grant was absorbed by the possibility—no longer taken literally—that the animal and plant kingdoms were in fact not separate and distinct from each other, but that certain primitive organisms like sponges and flustrae had some of the properties of both.

Darwin observed and collected lower animals on the seashore with Grant as early as February 1826. His diary entry for February 15 states that he "caught an orange coloured globular (Zoophyte) [that] was fixed to a rock & when kept in a bason would turn itself inside out, & when touched retracted itself in again; much in the same way as a Glove is turned inside out; put in spirits."[5] (Possibly this was an anemone.)

Two "philosophical men" had met, with Darwin as the eager pupil-amanuensis and Grant the earnest teacher and guide. In Grant, Darwin had discovered a possible new mentor, a substitute for Erasmus. In many ways, Grant was almost the direct opposite of Darwin's brother Erasmus, unpolished instead of refined. But they had a lot in common. Grant was a keen walker, and soon they were taking long hikes together in the country. Grant was also a great talker and would regale Darwin, as they walked, with tales of his multiple crossings of the Alps and other European travels. Grant had acquired a student both to teach and to share ideas with. He had had few such students; one of them had been Darwin's friend John Coldstream, who may well have been instrumental in bringing these two rather shy people together.

As had been the case for Darwin and his brother, a favorite walking route took them to the fishing ports of Newhaven and Leith, with their herring and oyster fleets. They cultivated the fishermen and got good specimens in return. At other times, they combed the mudflats of the Forth estuary for specimens. Back in Edinburgh, they took their finds to Grant's rooms and there Darwin worked alongside Grant in his laboratory, using an old borrowed microscope. Evidently, Darwin did not already own a microscope; his experiences to this point had all been those of the field naturalist, focusing (so to speak) on the macroscopic world of hand specimens, whether minerals or dead birds. This was a new venture for Darwin, and he did not know where it would lead.

Under Grant's influence, Darwin developed into a more scholarly naturalist rather than a mere collector.[6] Grant showed him a direct connection between his zeal for the outdoors and the life of the mind. His physical energies were channeled into the field excursions along the coast to collect specimens and his mental energies into examining them minutely. Darwin later stated in his *Autobiography* that one of his

methods was always to make a note of something he did *not* understand it, as was often much more useful than a note on something that
he did.[7] The unfamiliar, the unusual, the familiar—all fascinated him.
He rarely made notes about things he already knew, trusting instead
to his memory and his capacity to make associations among known
phenomena.

Darwin had, of course, taken out a ticket for the library again (having
reclaimed his deposit the previous April).[8] That he kept up his program
of reading is demonstrated in a list that has survived in which Darwin
wrote down: "*Books that I have read since my return to Edinburgh.*"[9]
Medical works were definitely in the minority, although they were included: Volume 1 of John Abernethy's *Physiological Lectures* (1817) and
his *Hunterian Lecture and Oration* (1817), John Ayrton Paris's *Phar-
macologia* (1820), John Bostock's *An Elementary System of Physiology*
(1828), and William Henry's *The Elements of Experimental Chemistry*
(probably the ninth edition, 1823). More general works he read included Hugh Blair's *Lectures on Rhetoric and Belles Lettres* (1783), H.
K. White's *Letters and Poems* (1807), "several essays in Rambler," and
Cuvier's *Essay of the Theory of the Earth* (the 1827 edition, which he
bought).

Echoing his early days at school, this list is especially strong in
books on travel and exploration. At that time, not only had explorers all
over Europe pushed eastward and southward across the Old World,
there were the adventures of Lewis and Clark in the New World, published in England for the first time in 1815. And it was an especially exciting time for exploration of the Arctic. A new search for a Northwest
Passage had been taken on by the British navy. Sir John Franklin's epic
journeys on land across northern Canada had caught the whole nation's
imagination. Darwin read Sir John Franklin's *Narrative of a Journey to
the Shores of the Polar Sea* (1823), William Scoresby's *Account of the
("Polar") Arctic Regions* (1820), and Thomas Pennant's *Arctic Zoology*
(1784). Also on his list were "Clark's Travels" (possibly this was
Edouard-Daniel Clarke, *Voyages en Russie, en Tartarie et en Turque*

[1813]) and Charles Cochrane's *Journal of a Residence and Travels in Colombia during the Years 1823 and 1824*. Letters from these explorers and extracts from their works were regularly published in Jameson's *Edinburgh New Philosophical Journal*.

Significantly, Darwin had also extended his voluminous reading to the original scientific literature. In the same notes, he mentioned having read "Pamphlets by Drs Grant and Brewster on Natural History," "Several papers in the Wernerian Trans.," and "Several numbers in ye New Edinburgh Philos. Journal." Some of these would have been recommended by Jameson during his lectures, of course.

It was during this second year at Edinburgh that he bought and read his grandfather Erasmus Darwin's great treatise *Zoonomia; or, The Organic Laws of Life* (1794), although he may have read it earlier as well. That Charles Darwin was intensely interested in his grandfather's ideas at this time is shown by the fact that he also mentioned having read Anna Seward's scandalous biography of Erasmus Darwin. (Toward the end of his life, Darwin wrote his own account of his grandfather's life that he appended to the English edition of a biography by Ernst Krause.)[10]

In this second year at university, and especially at meetings of the Plinian Society, Darwin experienced directly the pervading controversies of Edinburgh's intellectual life and its combative styles. Geology was not the only hot topic of the day in natural science. Closer to the hearts of medical men was the question, posed by Descartes, among others, concerning the material nature of the body. As physiologists and anatomists slowly pieced together (or rather took apart) knowledge of the functional relation between nerves and muscles and the role of electric impulses in the firing of nerve cells, and as they examined the brain, the obvious question arose: Is the mind anything more than the sum of activity in the nerve cells in the brain? Or is the mind something nonmaterial as well? And, of course, not just the mind but its transcendent sister, the soul—and, indeed, life itself?

That Darwin noted that he had read Abernethy's "Hunterian Lectures" may be a signal that he (and his teachers) had been following the dispute over this issue of the mind and matter. The great surgeon William Hunter had been one of the greatest defenders of the principle of vitalism, which propounded that the mind and body were animated by an immaterial "vital essence," rather than life being a "property of organization." Abernethy used the occasion of giving the Royal College of Surgeons' lecture in honor of Hunter to attack the London surgeon William Lawrence for his supposed adoption of dangerous (French) ideas about the physiology and chemistry of the body and the material basis of life. In reply, in a published series of lectures given at the Royal College of Surgeons, Lawrence sarcastically admitted that he had indeed been guilty of "not being fully convinced that the pretended Hunterian theory of life is the most important subject that can be entertained by the human mind. This slowness of belief must be pardoned."[11]

As for the mind, Lawrence averred, "the theological doctrine of the soul, and its separate existence, has nothing to do with this physiological question, but rests on a species of proof entirely different. These sublime dogmas could never have been brought to light by the labours of the anatomist and physiologist. An immaterial and spiritual being could not have been discovered amid the blood and filth of the dissecting-room." All this did not mean, of course, that either Abernethy or Lawrence thought chemistry had no place at all in the study of medicine and the functioning of organs: "A close alliance between the science of living nature and physics and chemistry, cannot fail to be mutually advantageous."[12] The problem, of course, was that all sides of the argument were juggling closely held opinions and only small nuggets of relevant hard data. As we shall see, exactly the same would be true in the case of the strongly contested field of "transmutation of species."

March 27, 1827, turned out to be a crucial date for Darwin. At the meeting of the Plinian Society on that day, Darwin's friend William Browne presented a paper on this exceptionally contentious subject of mind and matter. He had discussed the issue at previous meetings; now he came out with a full-blown treatment. His presentation, "On Organization as

Connected with Life & Mind," concluded that mind and consciousness were material. The meeting ended in an uproar of discussion among Grant, Binns, Greg, Ainsworth and Browne. Browne had gone too far, even for a student society; when the minutes of the society for that day were later edited, Browne's contribution was struck out (but, perhaps deliberately, it was done so that one could still read what had been written).

To see his friend Browne at the center of such a row would have been distressing enough for Darwin, but there was more. The first student to present at that same meeting of the Plinian Society had been Darwin himself. The Plinian Society meeting was just as fraught for Darwin as for Browne: both knew that they were going to be deliberately confrontational.

On the face of things, Darwin simply presented the results of some of his work with Grant on primitive animals of the Scottish seashore. At just eighteen years old, not only was Darwin working alongside Grant, he had been making discoveries in parallel. By March, he had achieved his first real scientific "hit" with his microscopic investigations. "I showed that little globular bodies [then sometimes called sea-peppercorns] which had been supposed to be the young state of [the seaweed] Fucus loreus were the egg cases of the worm-like *Pontobdella muricata* [a marine leech]." More dramatic was his second discovery. His later written account begins, typically, with a modest disclaimer that actually becomes a boast: "From not having had any regular practice in dissection, and from possessing only a wretched microscope my attempts were very poor. Nevertheless I made one interesting little discovery . . . that the so-called ova of Flustra had the power of independent movement by means of cilia, and were in fact larvae."[13] This had never been observed before. Not incidentally, it helped confirm the impossibility of one of Grant's longtime dreams: "flustrae" were simply animals, not something halfway between an animal and a plant. The minutes of the meeting record that Darwin was asked to "draw up an account of the facts and to lay it, together with specimens, before the Society next evening."[14]

Darwin's portion of the meeting ended with Grant taking the floor and detailing "a number of facts regarding the Natural History of the

Flustra."[15] This should have been a nice event, showcasing cooperation between student and teacher. Instead, it was exactly the opposite.

There is no known date for Darwin's first observation of the motile "ova" of the flustra carbacea, although on March 19, only eight days before his Plinian Society presentation, he noted in a notebook that he apparently started especially for this occasion: "Observed ova in Flustra Foliacea and Truncata, the former of which were in motion."[16] Grant's own work on these creatures had already filled some twenty scientific papers in the journals of the day. But in all his researches, Grant had not observed the motile stage. So, as soon as he first saw the ciliated larvae, Darwin wanted to show Grant his discovery. To his consternation, Grant's response was chilly. "He [Darwin] rushed to Prof. Grant who was working on the same subject to tell him, thinking, he wd. Be delighted with so curious a fact. But was confounded on being told that it was very unfair of him to work at Prof. G's subject and in fact that he shd take it ill if [he] published it."[17]

Grant then launched into a most unfriendly response. Not content with trying to upstage Darwin with his own superior knowledge of flustrae at the Plinian Society, three days before Darwin's scheduled presentation at that meeting he read a paper of his own— "Observations on the Structure and Nature of Flustrae"—to a meeting of the Wernerian Natural History Society. In it, he reported Darwin's discovery as if it had been his own and without acknowledgement. He also appropriated Darwin's discovery about the leech *Pontobdella*. It is not known whether Darwin was a guest for this Wernerian Society meeting.

In view of Grant's behavior, it was an act both of open defiance and self-protection on Darwin's part to go ahead with his own presentation three days later. It also required a great deal of courage. The depth of ill feeling shows up in the brief account Darwin later wrote of the discovery:

April 20th Having procured some specimens of the Flustra Carbacea (Linn.) from the dredge boats at Newhaven; I soon perceived without the aid of microscope small yellow bodies studded in different directions on it.—They were of an oval

shape & of the colour of the yolk of an egg, each occupying
one cell. Whilst in their cells I could perceive no motion; but
when left at rest in a watch glass or shaken they glided to & fro
with so rapid a motion as at some distance to be distinctly visi-
ble to the naked eye. When highly magnified the cilia, which
were chiefly distributed on the broader ends were seen in
rapid motion; the central ones being the longest. I may mention
that I have also observed ova of Flustra Foliacea and Truncata
in motion. That such ova had organs of motion does not appear
to have been hitherto observed either by Lamarck Cuvier Lam-
oureux or any other author:—This fact although at first it may
appear of little importance yet by adducing one more to the already numerous examples will tend to generalize the law that
the ova of all Zoophytes enjoy spontaneous motion. This and
the following communication was read both before the Wernerian & Plinian Societies.[18]

The "following communication" concerned his observations on the egg
masses of *Pontobdella,* previously thought to be those of Fucus. There
is no record in the minutes of the Wernerian Society that Darwin's paper was presented there. Perhaps it was done so unofficially. Perhaps
Darwin merely had had hopes of such a submission at the time he wrote
the notes.

There is something triumphant in Darwin's noting that the motion
of the "ova" could be seen with the naked eye but had not been observed by "any other author." Grant, the local expert, was carefully not
named. The two concise essays also show a quite mature style, reflecting his habits of work and his catholic reading. This was no casual amateur. In the space of a little more than two years from leaving Dr.
Butler's school, Darwin had acquired a considerable maturity and had
developed a pattern of working that would stand him in good stead for
the rest of his life.

When Grant later published his long Wernerian Society paper on
flustrae in the *Edinburgh New Philosophical Journal,* Darwin again
looked in vain for mention of his own contributions. Not only had

Grant tried to scoop Darwin in terms of the timing of the announcement, he had now taken all credit for himself. This account by Grant was very carefully worded: "By examining the ovum within [its] capsule, with the microscope, we perceive its ciliae in rapid motion; and I have frequently observed the ovum, in this situation, contract itself in different directions, shrink back in its capsule, and exhibit other signs of irritability before its final escape . . . on escaping from the cell, the ovum glides to and fro by the action of its ciliae, and, after fixing, it is converted to a single complete cell, from which new cells shoot forward."[19] In this, the words from "and I have frequently observed" to "escape" represented Grant's own observations. The rest was owed to Darwin.

Whether pressure was applied or not (it seems likely so, perhaps by Jameson), Grant finally made a grudging acknowledgement of Darwin's observations on *Pontobdella* in a later paper in the *Edinburgh Journal of Science*: "The merit of having first ascertained them to belong to that animal is due to my zealous young friend Mr. Charles Darwin of Shrewsbury, who kindly presented me with specimens of the ova exhibiting the animal in different stages of maturity."[20] But Grant never acknowledged Darwin's work on the more significant matter—the flustra. To complete this unhappy story, Darwin was soon to discover that Grant had previously treated Coldstream in an identically shabby fashion.

The period was the lowest point of Darwin's stay in Edinburgh. His first great adventure in science and firsthand exposure to the rough-and-tumble of the intellectual world of the 1820s had ended with a sour taste. One eagerly turns, therefore, to Darwin's *Autobiography* to see what he says about the unsavory affair of Grant and the flustrae. Oddly, he did not mention it. Evidently, even after the passage of fifty years, the wound was too great. But we know that Darwin was crushed by the denouement of the flustra researches because his daughter Henrietta later recalled: "This made a deep impression on my Father and he has always expressed the strongest contempt for all such little feelings—unworthy of seekers after truth."[21] Darwin's resulting distaste for Grant, once his respected mentor, is also shown in the statement in the *Autobiography* to the effect that Grant accomplished nothing in science when at London.[22] In fact Grant did a great deal.

In the *Autobiography*, what is not said is often as important as what is. That Darwin did not mention the affair with Grant, or even mention him much at all except to dismiss him, tells us, I believe, how deeply Darwin was wounded. But as we shall see, there are other, even more telling, reasons for Darwin to deny any scientific connection or debt to Grant.

Not only had Grant let him down badly as a personal mentor, Darwin was now thoroughly disenchanted with Edinburgh. Instead of enjoying the pleasures of the intellectual pursuit of natural science, he had found only controversy and betrayal. This, together with the feelings he had about the attractions of medicine as a career, confirmed his doubts about his commitment to his studies at Edinburgh. But one more issue concerning Robert Grant remains to be discussed. Grant was an evolutionist.

SEVEN

Lamarckians

In addition to debates over Neptunist and Vulcanist geology and the divine versus material source of the mind, the third controversial issue dividing colleague from colleague, and teacher from student, in Edinburgh was one that continues to be vexatious even today. It was the subject of evolution or, as it was known then, the transmutation of species. It is highly significant, therefore, that Jameson's lectures in his natural history course included at least two on "the Philosophy of Zoology," of which the first was "on the origin of the species of animals."[1]

Throughout the Enlightenment—the Age of Reason—the possibility of the transmutation of species one into another had been a subject in the background, and occasionally the foreground, of philosophers' minds. Robert Hooke, for example, had noted in 1694 that if extinction had been commonplace through the history of life, there had to be a parallel flood of origin of new species to replace them. "And it seems very absurd to conclude, that from the beginning things have continued in the same state that we now find them."[2]

If, because of extinctions, the world had constantly had to be repopulated, one possibility was that God recurrently created new species at new Creation events, just as he allowed others to become extinct. The more dangerous alternative was that (allowing for a divine initial creation of life) all species were descended one from another according to natural processes and laws. The most radical view, almost unmentionable in

71

the seventeenth century and most of the eighteenth, was that even the very first life had arisen without God's intervention.

A huge part of the philosophical revolution driven by Descartes, Hume, and Locke was a shift to the search for material causes of natural phenomena. Eighteenth-century "natural philosophy" (the greater part of what we now call science) focused heavily on the nature of the material world, with the discovery of natural laws and with the elucidation of material-driven processes. For many philosophers, John Locke among them, the sorts of miracles of which the Bible tells were either impossible or simply a misrepresentation of some natural phenomenon. In his posthumous *Dialogues concerning Natural Religion* (1779), David Hume put into the mouth of Cleanthes the assertion that the world is "nothing but one great machine, subdivided into an infinite number of lesser machines . . . all these machines, and even their most minute parts are adjusted to each other."[3] That machine was created by God but thereafter ran according to natural laws rather than divine interference. All this echoed the views, much earlier, of at least some classical philosophers, particularly the Epicureans.

Much of this inquiry, of course, ran counter to the theological view that "nature" was nothing more than a synonym for God and that, if nature's works and processes were a mystery, then that was because they were *his* works and processes. To probe—not too deeply—into nature was entirely pious; to probe so far as to substitute material processes and laws for God was blasphemy. It had been a time, therefore, of great debate within the vaguely defined fields that we now oppose as "science" and "religion."

Georges-Louis Leclerc, comte de Buffon, in Paris, author of the monumental *Histoire naturelle* (1749–1809), was perhaps the first to propose a "natural" mechanism by which new species might arise.[4] Like so many scholars after him, Buffon pondered over an essential phenomenon of all living systems: generation (reproduction, as opposed to vegetation, or growth). In "generation," whether the animal embryo starts out as a formless cell or (as was once thought) a complete animal in miniature, in either case, there appear to be—must be—some sort of instructions that tell each molecule making up the body "where to go" and "what to do." Some molecules derived from food will become nerves, others muscles, or liver. With no idea of what the controlling mechanism or mechanisms might be, Buffon coined the metaphor of an "internal mold" (*moule interior*) according to which life processes were directed. It is not hard to see the force of such a metaphor. Just as a potter uses a mold to create a plate, the body must have its own internal instructions for making itself. The same internal mold would control lifelong processes of vegetation such as growth or, for example, wound repair. Our current view of the role of genetics and DNA fits exactly this metaphor.

Buffon's inspiration was that if each species had its own internal mold, and if that internal mold were somehow modified, then a different organism would be formed and propagated in the next generations. The sorts of phenomena that might cause an internal mold to become modified would include external environmental conditions. If the internal mold changed only a little, a new variety or race might emerge. If it changed more, there would be a new species. Each would be fitted to the new conditions that caused the change. Buffon thought that such change would mostly, if not exclusively, consist of degeneration, and his "internal mold" was *only* a metaphor. Today we would call it a "black-box" mechanism (meaning that we have no idea what its contents are).

David Hume also promoted the "principle" of generation (which he ranked as one of four basic principles of life: reason, instinct, generation, vegetation) and tried to tease out the relationships and differences between order that is "natural" and order that is imposed in the design by the Almighty. For example, in the *Dialogues,* Demea asks, "What is this vegetation and generation, of which you talk? . . . Can you explain their operations, and anatomize that fine internal structure, on which they depend? . . . how can order spring from anything which perceives not that order which it bestows?" Philo replies: "A tree bestows order and organization on that tree, which springs from it, without knowing the order: An animal, in the same manner, on its offspring."[5]

In his books *Zoonomia* and *The Temple of Nature,* Darwin's grandfather Erasmus took the concept a little further than Buffon. He wrote directly of an origin of all life from nonlife, in the sea. And he proposed that all life has arisen by a process of continuous change from a single, very simple, ancestral "filament." Later, the term "monad" would be used for this ur-life form. In his theory, all life on earth, living and fossil, had arisen by processes causing change within the generative properties of this original filament.

Erasmus Darwin proposed several ways in which change in the black box controlling "generation" might be caused. In addition to the influence of the environment, change might be induced by breeding and by "use and disuse." The first of these was simple enough, although it was impossible to know how. The second was supported very strongly by the long-standing practices of farmers and others. No greater example could be given of the power of "artificial selection" than the varied races of dogs, from Pekinese to St. Bernard, produced within the last two thousand years. The third mechanism—"use and disuse"—was much more attractive but shot through with difficulties. In the hoary old example, "use and disuse" meant that generations of giraffes, reaching higher in trees for food, stretched their necks, and a longer neck then became enshrined in the generative properties of their modified "filament." A blacksmith's children would have powerful arms. Similarly, but in the opposite direction, an organ that was not used (the human appendix, for example) would degenerate.

Erasmus Darwin's ideas implied that all living and fossil organisms are related, although he decided that animals and plants might have had separate origins. Very soon after Erasmus Darwin published *Zoonomia*, Chevalier Jean-Baptiste de la Marck (Lamarck), a zoologist at the Muséum National d'Histoire Naturelle (the former Jardin du Roi), published his own, elaborated versions of the same ideas. The first appeared in his *Système des animaux sans vertèbres* (1801) and was amplified in his *Philosophie zoologique* of 1809. Larmarck took over Erasmus Darwin's ideas of "use and disuse" and developed it to the extent that we now generally associate the concept with his name as "Lamarckism."

In Lamarck's view, species did not become extinct but morphed into others. Lamarck also thought about the driving processes and adopted Erasmus Darwin's ideas, especially emphasizing the direct effects of a changing environment and the heritability of features modified by use and disuse. He extended this by advocating the power of the "will" to force change in a particular direction. This added a new layer of black-box mystery. Meanwhile, as lineages progressively "improved" and increased in complexity, as revealed by the increase of complexity seen in the fossil record, waves of spontaneous generation resupplied the base complement of "lower" organisms.

Around the turn of the century, other philosophers, all from continental Europe, proposed schemes of evolution in which they similarly tried to account for the progression of increasing complexity of life seen in the fossil record as some form of continuum. In some rival theories of organismal change, degeneration was *the* driving force, and that was theologically attractive because it implied that God's original Creation had been perfect and complete. Few scholars, however, fully embraced the concept that one species would actually give rise to another. But the only alternative view—that God had created each species specially, and that he had done so either continuously or episodically over a hugely long period of time—was itself falling out of favor. If God had been the author of all those Creation events, why had he gone to the trouble of creating different kinds of species in different parts of the world? Why no penguins in the Arctic or polar bears, why no hummingbirds in Europe? Did extinction mean that he was constantly changing his mind?

By Charles Darwin's time at Edinburgh, most scholars were familiar with the concept of transmutation of species, even if they did not know how it might work. Most did not believe it in the first place. It was hardly a popular concept because it ran counter to every Christian teaching. Nonetheless, it had become one of the materialist subjects spawned by the Age of Reason that would not go away.

Darwin owned a copy of Cuvier's *Essay on the Theory of the Earth,* the new fifth edition, edited by Jameson. In the preface Jameson had written, "Can it be maintained of Geology, which discloses to us the history of the first origin of organic beings, and traces their gradual development from the monade to man himself,—which enumerates and describes the changes that plants, animals, and minerals—the atmosphere, and the waters of the globe—have undergone from the earliest geological periods up to our own time, and which instructs us in the earliest history of the human species,—that it offers no gratification to the philosopher?" This is much stronger language than that he had used in his 1813 edition, where he wrote: "To the geologist, this beautiful branch of Natural History . . . points out the gradual succession in the formation of animals, from the almost primeval coral near the primitive strata, through all the wonderful variety of form and structure observed in shells, fishes, amphibious animals, and birds, to the perfect quadruped of alluvial land."[6] There was no mention there of monads and no connection to humans there. Something had changed in between, and *that* was a surge of interest in the ideas of Lamarck and, by association, Erasmus Darwin.

In the *Autobiography,* Darwin noted that during one of their long walks in the country, Grant, a man who was "dry and formal in manner, but with much enthusiasm beneath this outer crust . . . burst forth in high admiration of Lamarck and his views on evolution." Given the topicality of Lamarck's theory, local connections to Erasmus Darwin, and Grant's period of study in Paris, his enthusiasm for the subject is hardly surprising. And who would blame Grant for launching into a discussion of his enthusiasm for Lamarck's ideas when he was walking along and talking about science with the great Erasmus Darwin's own grandson? Darwin's stated response, however, is most puzzling. "I listened in silent astonishment, and as far as I can judge without any effect on my mind. I had previously read the *Zoonomia* of my grandfather, in which similar views are maintained, but without producing any effect on me."[7]

The account of this episode with Grant has become one of the most quoted passages in Darwin's *Autobiography.* However, just why Darwin was astonished is not easy to see. What was Darwin really telling us? Was it Grant's Lamarckianism that astonished him, or the vehemence of his feelings, or the fact that Grant would confide such radical views to him? Was it that Grant thought his ideas were radical at all?

Is Darwin telling us that he had never heard of Lamarck before? Given that Darwin's account of Jameson's Neptunism was at best disingenuous, we must probe deeper to discover what this passage means.

Very many authors have taken Darwin's words at face value and interpreted them as direct evidence that Darwin had never really thought about evolution at that point, and—perhaps even more significant—that this chance encounter with the subject had no effect on him. If that were the case, Darwin's own later evolutionism could only be seen as something of his own devising, owing nothing to people or events in Edinburgh. As Darwin coyly put it, at the very most—"It is probable that the hearing rather early in life such views maintained and praised may have favoured my upholding them under a different form in my *Origin of Species*. At this time I greatly admired the Zoonomia; but on rereading it a second time after an interval of ten to fifteen years, I was much disappointed; the proportion of speculation being so large to the facts given."[8]

That Darwin would have been in any way astonished by Lamarckian views per se seems unlikely. After all, this was the grandson of Erasmus, who had been reading *Zoonomia* that very year. He had been more than a year in Edinburgh, which was a hotbed of every revolutionary and many reactionary ideas, especially those from Europe. Grant had just published a paper (April 1826), on the sponge genus *Spongilla,* with the rather modestly evolutionary view: "Descendants [of the early members] have greatly improved their organization, during the many changes that have taken place in the composition of the ocean. . . . it is highly probable that the vast abyss, in which the spongilla originated and left its progeny, was fresh, and had gradually become saline."[9] This was Grant's first public expression of an evolutionary viewpoint. Grant seems to have given Darwin copies of his works; it seems likely that he would have been particularly careful to give him a copy of this paper.

Under the direction of either Jameson or Grant, Darwin had used a copy of Lamarck's essay of 1801, in the preface to which Lamarck first outlined his evolutionary views.[10] He had even copied out Lamarck's classification scheme from that paper. But there is even stronger evidence to suggest that Darwin *must* have had direct exposure to Lamarck's ideas at Edinburgh. In October 1826, there appeared an anonymous article in the *Edinburgh New Philosophical Journal,* entitled "Observations on the Nature and Importance of Geology." This paper, which appeared in the same issue as Benoiston de Chateaubriand's essay on population dynamics, is a succinct exposition and critique of a Lamarckian evolutionary view. In fact, on page 297 it uses the word *evolve* in a transmutationist sense—the first such use in the English language: "Various forms have been evolved from a primitive model . . . species have arisen from an original generic form."[11]

Many scholarly pages have been devoted to finding the author of this paper. But it is pretty obvious that it was written by Jameson. It has just his schoolmasterish style, and the whole thing is framed exactly like one of the "Geological Illustrations" that he added to his editions of Cuvier's *Essay on the Theory of the Earth.* James Secord suggests that the essay was in fact something that Jameson had originally intended to be included in the fifth edition.[12] Perhaps Jameson refrained from publishing it there out of sensibility to Cuvier because the ideas expressed in the essay are essentially Lamarckian and espouse a view of life evolving gradually and continuously, opposite to Cuvier's views. In the 1827 edition of Cuvier, Jameson did include a fairly heavy criticism of Cuvier's catastrophist ideas in a twenty-page essay, "On the Universal Deluge." But he drew short of promoting Lamarck's theories in that essay.

The anonymous essay in the *Edinburgh New Philosophical Journal* began by extolling the importance of geology in discovering not only the history of the earth but also life on it: "If we contemplate its [the earth's] surface, with all its inequalities, it is geology alone that can give us a distinct representation of them. All local descriptions, not springing from this source, either leave behind them indistinct and erroneous conceptions, or are entirely fanciful. . . . [Facts concur with] historical testimony, in representing the elevated platforms of Asia as the cradle of the human race, and in explaining their diffusion from that centre; and the traditions of deluges, found among all nations of antiquity, are corroborated by the still existing traces of those violent events. . . . The monuments concealed in the bosom of the earth, and

extending to the whole organic creation, are still more instructive."
(This is very Cuvierian.)

The author pointed out that while the primitive rocks of the earth
are more complex than the newer ones, "in complete opposition to this,
the organic world, in each of its two principal divisions, exhibits a series of formations from simple to compound; the simplest being the
oldest. Thus we observe animal life commencing in infusory animals
[protozoans], without any discernible organs." As organs appear, "generation preserves the peculiar character of organic beings. . . . Have
they originated in the way in which they appear in the scale of gradation, as if the hand of the Creator, like that of a human artist, perhaps,
must first be exercised on simple formations, before it was capable of
producing such as were compound? Upon these questions, whose answer might contain no less than a key to the profoundest secrets of nature, Mr Lamarck, one of the most sagacious naturalists of our day, has
expressed himself in the most unambiguous manner. He admits the
existence of the simplest worms, by means of spontaneous generation,
that is, by an aggregation process of animal elements; and maintains,
that all other animals, by the operation of external circumstances, are
evolved from these in a double series, and in a gradual manner."

For the anonymous author, "geology alone can inform us, how far
this successive course of development might have been followed by nature." And then he follows Cuvier and departs from Lamarck in stating: "Geology does not inform us merely of the origin of animal
species, but also of their destruction. Out of the vast number of animal
remains, but few belong to species now living, and these only, in the
most recent rock formations; by far the greatest number of their primitive structures were lost."

Then he picked up another theme from Lamarck—and, of course,
from Erasmus Darwin—before allowing that Lamarck might have been
partially correct on extinction. "The distinction of species is undoubtedly one of the foundations of natural history, and her character is the
propagation of similar forms. But are these forms as immutable as some
distinguished naturalists maintain; or do not our domestic animals and
our cultivated or artificial plants prove the contrary? If these, by change

of situation, of climate, of nourishment, and by every other circum-
stance that operates upon them, can change their relations, it is possible
that many fossil species to which no originals can be found, may not be
extinct, but have gradually passed into others."

Finally, "If [changes in the organic world] have required intervals
of time that are antecedent to all historical transitions, and to the duration even of the human race, the monuments concealed in the bosom of
the earth can alone reveal them. We indeed observe that the Ibis, which
was worshipped in ancient Egypt, and preserved as a mummy, is still
the same in modern Egypt; but what are the few thousand years to
which the mummy refers, in comparison with the age of the world, as
its history is related by geology."

It has been thought that, alternatively, Grant, as the prime Lamarckian in Edinburgh, might have been the author of this paper. Its subject, however, was geology, and that was not Grant's forte. Grant could
never have written the words "Geology alone can inform us." That the
author was Jameson seems confirmed by the appearance of a second,
less often noticed, anonymous article. This appeared in the same journal with the title "Of the Changes Which Life Has Experienced on the
Globe."[13] It also fits the style that Jameson used for his "Geological Illustrations" in the Cuvier work, and it again adopts a Larmarckian approach, specifically arguing that life on earth has changed gradually
rather than episodically and under the influence of a changing environment. This second essay was contained in the same issue of the
journal as the concluding part of Grant's flustrae paper. Its wording is
very interesting.

"The beings, which were unable to resist the influence of these various causes were destroyed and disappeared from the earth, with the
circumstances for which they were created; new species appeared with
new conditions of existence." And again, the essay specifically counters
Cuvier's catastrophist view of earth history. "But, in examining the series of fossil remains that are found buried in the strata of the globe,
there is nowhere perceived a distinct line of demarcation between the
different terms of that series, so as to prove that life has been once or often totally renewed on the earth. On the contrary, we discover in it a

proof of the successive and gradual change which we have pointed out. Certain primitive types have indeed completely disappeared, but they are found existing at various epochs, and their remains are blended with those of more modern types; along with new species of types still existing, we find some of anterior epochs."

Humans, however, presented a problem. If life is constantly undergoing change, will humans in turn be eliminated? The first anonymous paper avoided the subject. In the second essay the author's view is the orthodox one that all in the earth's history was pointing to the arrival of humanity, and humans heralded the end of change. "Man appears to have arrived upon the earth only after its surface was adapted to receive him, after the establishment of climates, and when a happy equilibrium among the elements had determined the permanency of the present state of things, or at least had rendered its variations almost imperceptible."

Taken together, these two articles fit the pattern of Jameson's "Geological Illustrations" perfectly. However, a more remote possibility exists, that the author of the second anonymous paper, which repeats the Lamarckian theme of gradual change expressed in the first, was someone who had felt piqued that Jameson had published the first essay and stolen his own evolutionary thunder. If this scenario is correct, then the author of the second paper might have been Grant, after all. Perhaps Grant wanted to put down a marker of his own.

The likeliest interpretation is that Grant wrote neither paper. Simply, on the famous occasion of his discourse with Darwin, Grant was angry that he had not had the courage yet to write such articles as these—*that* might indeed have provoked an outburst and it might indeed have been an astonishing one.

The *Edinburgh New Philosophical Journal* was something that academics in Edinburgh read avidly, especially because Jameson included much controversial material along with bulletins from explorers. Darwin's intimate familiarity with the journal in which the two papers appeared is shown by the fact that he owned copies of at least some volumes. While preparing for the *Beagle* voyage in September 1831, Darwin wrote to his sister Susan, asking her to look on his bookshelf papers for "the Edinburgh Journal of Science, or some such title, & see whether the following papers are in it: 3 by Humboldt on isothermal lines: 2 by Coldstream & Foggo.—on Metereology: Metereological observations."[14] We know that Darwin also had reprints of Grant's papers from the *Edinburgh Philosophical Journal* (given him by Grant). They still exist in the Darwin papers at Cambridge. Coldstream was also his friend and probably also gave him copies of his work. But to have owned the Humboldt papers, Darwin had to have bought or been given the journals.

If, as a student in Edinburgh, Darwin had been interested in subjects as disparate as climatological readings in Seringapatam (Shrirangapattana), India, and the small fishing town of Leith in Scotland, plus Humboldt's more intellectually exciting ideas on isothermal lines and their relationship to patterns of vegetation in South America and Europe, *and* also remembered the citations four years later, then it seems reasonable to assume that he also had had every opportunity to read the first of the two anonymous papers on transmutation, if not both.

Given the content of the two Lamarckian papers published in the *Edinburgh New Philosophical Journal,* it is inconceivable that Darwin and his circle had been ignorant of them and or that they failed to discuss them, especially the first one. One can hardly imagine the members of the Wernerian and Plinian societies, or groups of members gathering informally, finding the subject matter uninteresting and noncontroversial.[15]

In the end, however, the authorship of these two papers probably does not matter. And, even if we were to ignore the question of whether Darwin actually read them, the fact remains that two Lamarckians were active in Edinburgh in 1826–27. Both men—Jameson and Grant—were teachers of Darwin. In this context, the fact that Jameson ended his lecture courses with a discussion of transmutation of species becomes more important. And all of this makes Darwin's expressed "astonishment" itself astonishing.

We will never know the exact details. However, Darwin's profession of astonishment at Grant and his Lamarckian outburst makes one worry that here, as in so many other matters, his *Autobiography*'s misdirections and forgetfulness may be rather convenient. Even while he enjoyed a well-earned, heralded career in his old age, Darwin was incapable of sharing credit, of finding the shades and subtleties of intellectual debt that any creative person knows. It seems a curious and terribly damaging weakness to be so unsure of one's own accomplishments. But then we also have to remember that the *Autobiography* was written in 1876 when Darwin was famous for evolution, Lamarck's ideas were scientifically passé (although they had a popular following), and Grant was notorious for being a radical. It would have been hard for Darwin to find a way to admit that he had at any time been intellectually close to, or dependent on, either of them.

From November 1826 to April 1827, Darwin had led a tripartite life. Jameson's disciplined and rigidly organized course occupied his classroom and museum hours; Grant and the Plinians provided a world of conflicting ideas and the attractions of independent laboratory work; meanwhile, both of these were set against the insistent demands of increasingly unattractive medical studies. Darwin probably revealed so little of all this in his *Autobiography* because he could write his life only seen through the prism of his own work which, at this stage, he had not yet defined in *his* terms.

If this reading of the *Autobiography* seems unflattering to Darwin, the alternative is really absurd. It would mean that not only did Darwin mysteriously acquire a considerable knowledge of geology by the time of the *Beagle* voyage, but he had also managed to forget a great deal of exposure to the theories of transmutation that were then circulating in intellectual milieus. But it is not necessary, in order for us to accept that Darwin was a genius, to believe that he was also totally self-taught or that he created his ideas out of nothing. By the end of this book, I will have presented other examples showing that Darwin created a great deal out of existing ideas and concepts. And the greatness of his contributions will, in the process, be enhanced rather than diminished. That greatness lay in analytical thinking, rather than intuition, scientific rather than artistic judgment, and painstaking observation rather than fiery debate.

EIGHT

Cambridge Undergraduate

Darwin left Edinburgh in April 1827 planning to spend the summer as he had previous summers, "wholly given up to amusements, though I always had some book in hand, which I read with interest. . . . the autumns were devoted to shooting, chiefly at Mr. Owen's at Woodhouse, and at my Uncle Jo's, at Maer. I kept an exact record of every bird which I shot throughout the whole season."[1] This was the summer when Darwin finally seems to have discovered an interest in the opposite sex.

The Owen family were close neighbors of the Darwins. They kept a jolly, crowded, household, rather in contrast to the somber mood at the Mount: "It is hardly possible for common mortals . . . to wind up their spirits to the Woodhouse pitch." There were four daughters, great friends of Darwin's sisters. Darwin became entranced by Fanny Owen, then aged nineteen and "the prettiest, plumpest charming Personage that Shropshire possesses."[2]

It had been another idyllic summer, but life was getting complicated. There is no record that Darwin worked with his father in his medical practice, and he seems to have done everything he could to avoid being at home in Shrewsbury. It was not just the charms of the Owen girls that kept Darwin away from home. Once the autumn shooting had come and gone, the date for return to Edinburgh approached fast. But Darwin had decided that he would not go back to Edinburgh and would not continue to study medicine.

Just when Darwin made these two decisions is not clear. There is also a serious question about just *why* he decided against continuing at Edinburgh. As already noted, most authors cite as the turning point his horror at the surgeries he witnessed. But that had happened in the previous year. The telling event of 1827 was his betrayal by Grant in the matter of flustrae. Up to that point, he seems to have been thoroughly enjoying his immersion in natural science, if not in medicine. Edinburgh—with its rivalries and intense intellectual debate—was too raw and robust for his shy and retiring disposition. He had seen his friends savaged at the Plinian Society and wanted to retire from a scene in which the intellectual life was so personal. He may also have been depressed that Jameson's course—whether or not he had learned much from it—had left him unnerved even about his favorite subjects in natural history. In addition, even if things had been patched up with Grant, the latter was leaving for the University of London. Darwin would once again be without a mentor.

With all its bad associations, Edinburgh was no place for him to continue in a subject that he disliked. He needed to regroup, even though that was bound to cause his father to be intensely disappointed in him.

At some point, Erasmus had told Darwin the financial facts of life, revealing the tantalizing truth that money was never going to be a problem. Both young men were amply provided for in their mother's will. The Wedgwood money would come to them at age twenty-one. Their father already supported them generously; when he died, they would be wealthier still. That news probably did a lot to discourage Darwin from continuing to persevere in a profession he hated. But it had not been a reason for Erasmus to give up the study of medicine. A Darwin needed a profession, after all. Erasmus went on to qualify but never practiced medicine, instead living an elegant and undemanding life as a literary hanger-on in London society.

All through the summer of 1827, Darwin failed to screw up the courage to discuss the problem with his father. Eventually, "my father perceived or he heard from my sisters, that I did not like the thought of being a physician."[3] One can only guess what torrents of recriminations this produced. This may have been when Robert Darwin uttered his memorable condemnation: "You care for nothing but shooting, dogs, and rat-catching, and you will be a disgrace to yourself and all your family."[4]

Darwin might secretly have preferred to avoid university altogether and simply become a gentleman naturalist. However, if taking up a career was obligatory and medicine was out of the question, the remaining choices, for a family like the Darwins, were the army, law, or the church.[5] Darwin's nervous disposition would have failed completely under army disciplines and pressures. The law had not been a happy choice for Darwin's uncle. With no other choices, "[my father] proposed that I should become a clergyman."[6] This all seems a bit cold-blooded, lacking in passion or even simple enthusiasm. And the church presented all sorts of new problems. First and foremost, except for a practiced cynic, something of which Darwin was never accused of being, going into the church required all sorts of commitments, far greater commitments in a moral and spiritual sense, than training for medicine. It is the one career where Talleyrand's injunction "Surtout, pas de zèle" should not apply. The church, moreover, was a public life, and Darwin was dreadfully shy.

On the positive side, should he get through the obstacles, the life of an affluent country parson would be ideal in allowing plenty of time for natural history pursuits—the model of the Reverend Gilbert White, author of *The Natural History of Selborne,* had already attracted many interesting young men to a career in the church. If he could obtain the ideal sinecure, one of those parishes (*livings* was such a perfect term), he would have to turn up only a few times a year to minister to his flock. Even that, however, would require considerable development of his father's gift of chatting up old ladies.

Over all, there hung the issue of the subject matter—to enter the church one had to *believe,* not exactly a minor point. There was no hint of a vocation in Darwin's decision to follow his father's orders. The family had slid from the serious Unitarianism of Erasmus Darwin and of Charles Darwin's late mother to the socially acceptable Anglicanism of the Christmas-Easter-weddings-funerals kind fitting Dr. Darwin's status in county society. It would be nice to know that Darwin or his father consulted the vicar of St. Chad's. However, we do know that Charles Darwin took the religious part of the decision seriously. He must have been acutely aware that he was not cut out to be a parson, starting with the problem that his belief in things Christian was of the usual, rather perfunctory, kind instead of the deep personal conviction that any vicar (literally a representative of God) should have.

His approach to the problem was typically academic. He probed into the extent of his beliefs. Simply to matriculate at Cambridge, he would have to subscribe to the "Thirty-nine Articles" of the Anglican Church. He tells us that he settled down and made a critical reading of an old classic: Rev. John Pearson's *Exposition of the Creed*, setting himself a test—if Pearson made sense, he could do it. Somewhat to our surprise—but not, one suspects, to his—he found the rhetoric convincing. It followed the familiar argument of conventional belief: "As I did not in the least way doubt the strict and literal truth of any word in the Bible, I soon persuaded myself that our Creed must be fully accepted."[7]

If this entry to preparation for the church was somewhat dispassionate and perhaps even noncommittal, it was not cynical. Possibly Darwin approached his time at Cambridge hoping that he would never have to be called to holy orders, just as Erasmus hoped never to practice medicine. Certainly, there was no question of his being called to minister to the poor of London, or Shrewsbury, for that matter. In the meantime, three years at Cambridge would give him something respectable to do. It would place him nicely in English society, meeting all the right people.

One other factor undoubtedly added to Darwin's willingness to go along with this new plan. Predictably, it involved his brother Erasmus, who would be at Cambridge too, at least until he passed his medical examinations in the spring. Without Erasmus there, Darwin might never have been able to tackle another new situation. With Erasmus present, Cambridge would be far less threatening than if he were to go alone; less threatening than another year at Edinburgh. Without Erasmus, it would have been well-nigh impossible. Another factor was that

Cambridge was not the place of intense intellectual turmoil that Edinburgh was. Study was focused on reading rather than lectures. Nor was the curriculum particularly demanding. He would have both the time and the intellectual "space" to indulge his passion for natural history.

In the end, he could not start immediately. For once, there actually were requirements to be met. He had to stay at home for the rest of the autumn in order to be tutored in the classics, which he had largely forgotten, and he matriculated at Christ's College, Cambridge, only in January 1828. By then, it would be only a few months before Erasmus left, heading for London to continue his studies and leaving Darwin once again isolated and alone. Worse, because of having matriculated at Christmas, Darwin was living not in college but in rooms on Sydney Street.

For the shy, undirected Darwin, these perturbations to an already shaky plan could have been disastrous, but luckily help was at hand in the form of his second cousin "Fox"—William Darwin Fox, a great-nephew of old Dr. Erasmus Darwin. Darwin had not known Fox very well up to this point, but Fox was also up at Christ's and Darwin now attached himself to him. He walked with Fox, shared the same interests in natural history, and attended Fox's teachers. Judging from Darwin's later correspondence with Fox, they had a number of friends in common, but they were mostly older men who graduated with Fox.

Fox, a very keen entomologist, was preparing for the church just as Darwin was, and had plotted out his career. He had the same career plan as Darwin. He intended to become a serious naturalist while taking care of an undemanding church in the country. In this he was successful, spending his life as the Reverend Fox, rector of Delamere in Cheshire, where he was occupied with wife, children, and collecting insects. In all ways, Fox was an admirable substitute for Erasmus and just the mentor Darwin needed.—"I live almost entirely with Fox," he wrote to Erasmus.[8]

To be at Cambridge in 1828 was not to partake of one of the shining moments of the English university system. It was possible to be a serious scholar there, but most undergraduates were what today would be called "slackers." For those students who did not intend to try for "honors," it was more of a country club than a place of learning. Fox was no scholar.

But neither he nor Darwin was interested in a dissipated life either. Field sports and natural history filled their days. Darwin kept a horse at Cambridge and rode both for pleasure and to get out to distant field sites to collect insects.

In company with Fox, Darwin became the well-heeled amateur naturalist/sportsman and all-round sporting gentleman. Together, Fox and Darwin joined in the great new craze of "beetlemania," for which they were both already eminently qualified. One great advantage of beetle collecting, as opposed to, say, butterflies, is that it could be pursued across the fenlands through the raw, cold winter and spring months, by digging into rotten logs, under stones, and so on.

Typically, Darwin threw himself into this with complete seriousness. One well-known story has him so intent on catching yet one more specimen, although his hands were already full, that he popped a beetle into his mouth—with the predictable result. This was not an idle hobby; it was almost a competitive sport—and not just a sport, because the mania for assembling a collection extended to purchasing specimens from commercial collectors. It was an all-consuming occupation, with the aim of assembling as complete as possible a set of British species. The rare species were obviously the most prized.

That summer, Darwin's entomological aspirations were reinforced by a meeting with one of Britain's great insect collectors, the Reverend Frederick William Hope. Hope was a wealthy man who collected art as well as insects; he donated his collections and a professorship to Oxford University. Darwin wrote in great excitement to Fox, "Now for the glorious news. I have been introduced, & if I may presume to say so, struck up a friendship with Mr Hope. I met him at dinner, & I find he knows all my Scotch friends, & we had so much entomological talk, that he asked me to bring over all my insects. . . . My head is quite full of entomology. . . . He thinks he can give me 3 or 400 species at Christmas. . . . I could write all day about him. . . . he has given me a great many water beetles."[9]

These happy days continued until January 1829, when Fox went up to Cambridge early to sit his final examinations. Darwin, still at home for the tail end of the vacation, wrote in support: "From the bottom of my heart I pity your dismal state in Cambridge: I suppose there is . . . this odious weather."[10]

Fox graduated—with some difficulty, perhaps understandable given the ways in which study had given way to entomologizing—and Darwin was left alone once again. Soon he underwent a distinct decline and neglected his studies even further.

The months following Fox's departure were the lowest point in Darwin's life at Cambridge. He had failed successively at Dr. Butler's school and at Edinburgh; now his world at Cambridge was falling apart. He had no close friends, no particular ambition, no interest in his academic studies—no reason for being at Cambridge except to prepare for a profession for which he had neither interest nor ambition. Insect collecting was a lot less fun when done alone. Whenever pressure built up like that, Darwin's health suffered: he became even more inward and gloomy, his skin problems flared up, his stomach and head ached, and he retired to his room. He wrote to Fox in April 1829, "If you did but know how often I think of you & and how often I regret your absence. . . . I find Cambridge rather stupid, & as I know scarcely anyone that walks, & this joined with my lips not being quite so well, has reduced me to a sort of Hybernation." Nonetheless he added, rather inconsiderately, "Old Whitley has begun to take your place."[11]

One salvation was that Darwin continued with his beetle collecting for the rapidly growing joint collection with Fox, although the letter just quoted continues: "Entomology goes on but poorly: a few Dromius & Agonum's, together with the Pecilus (with red thighs) make the g. part of what I have collected this term." Part of the therapeutic value of collecting was that it brought out his most competitive side. Among other students at Cambridge was a man whose passion for the insects was demonstrated in his soubriquet of "Beetles" Babington. When it turned out that both were buying specimens from the same dealer in London and that he was giving Babington first choice of the insects coming in, Darwin exploded. He wrote to Fox, "I have caught Mr. Harbour letting Babington have the first pick of the beetles; accordingly we have made our final adieus, my part in the affecting scene consisted of telling him that he was a d——d rascal, & signifying I should kick him down the stairs if he ever appeared in my rooms again: it seemed altogether mightily to surprise the young gentleman."[12]

One indicator of the seriousness with which Darwin pursued his entomology collecting, and the status that he was attaining, was that in February 1829 he went to London and spent two days with Frederick Hope. They "did little else but talk & look at insects: his collection is most magnificent & he himself is the most generous of Entomologists; he has given me about 160 new species, & actually often wanted to give me the rarest insects of which he had only two specimens; He made many civil speeches, & hoped you [Fox] will call on him some time with me. . . . He greatly compliments our exertions in Entomology, & says we have taken a wonderfully great number of good insects." A few days later, he "Drank tea" with James Francis Stephens, another major London collector and author of *A Systematic Catalogue of British Insects,* a multivolume work on which Darwin greatly relied. "His cabinet is more magnificent than the most zealous Entomologist could dream of."[13]

Stephens also published an eleven-volume *Illustrations of British Entomology,* which became an important reference for Darwin.[14] With the application of Linnaean principles of taxonomy and classification, these works ensured that entomology passed from simple collecting to a well-organized science. Even more important, in many ways, was *Horae Entomologicae; or, Essays on the Annulose Animals,* published between 1819 and 1821. This was a two-volume work by William Sharp MacLeay, an officer in the Foreign Service, secretary of the Linnean Society of London and a theoretical zoologist. At first Darwin probably knew of MacLeay's work simply as an adjunct to his insect collecting; later it became an important element in his search for a theory of evolution.

Insect collecting evidently brought out a strongly competitive aspect of Darwin's personality. Together he and Fox gossiped in letters about rival collectors like the Reverend Leonard Jenyns, a naturalist and author of great accomplishment with many Cambridge connections. In every respect he was the ideal example of the clergyman-naturalist that

Fox and Darwin aspired to become. But in letters to Fox Darwin disparaged him: "I am glad to find that neither [Stephens] nor Mr Hope have much opinion of Mr Jenyns; they seem to think him very selfish and illiberal."[15] Eventually they became great friends, and Jenyns was partly responsible for Darwin being invited to take part in the *Beagle* voyage.

We should perhaps be on our guard about the image that Darwin, in the *Autobiography,* tries to give us of himself during this period: "I got into a sporting set, including some dissipated low-minded young men."[16] He seems more depressed than dissipated. Perhaps this admission was intended to show him as only human after all: sowing his wild oats even if they turned out only to be clover. In his whole life Darwin was never a lighthearted sort of person. Pleasure for the sake of pleasure probably never occurred to him—the closest would have been his passion for shooting, but even there everything was meticulously accounted for in notebooks and diaries.

Reading is another answer to the apparent emptiness of this period, for the one word that one never associates with Darwin is lazy. He said that he read Locke (presumably *An Essay on Human Understanding,* which was part of his required study) and Adam Smith (whose works were not). He was unfocused and depressed but never unoccupied. Therefore we have to assume that, whether or not he was actually studying, Darwin, self-absorbed and intense as ever, read extensively, as he always had and always would.

Nonetheless, as far as formal studies were concerned, from January to October 1829 Darwin was drifting academically, as he always seemed to do when abandoned to his own devices. He made so little progress that Dr. John Graham, his tutor at Christ's, advised him not to go in for the Previous Examination ("Little Go") required of students at the beginning of their fifth term, but to put it off for a year. Given the fact that he had started at Christ's only in January 1828, he would have been taking the Little Go after only four terms of study, so perhaps the blame was not entirely on Darwin's side. Significantly, worrying about the examination, and also the fact that at Woodhouse he had accidentally shot one of the Owen boys over the eye, that January Darwin had the bad case of dermatitis of the lips that, as previously noted, prevented him from making a trip to Edinburgh to see his friends.

In the summer, he started out on an insect-collecting trip to Wales with Hope, but it was aborted—something Darwin would not have done casually. "The first two days I went on pretty well, taking several good insects, but for the rest of that week, my lips became suddenly so bad, & I myself not very well, that I was unable to leave the room, & on the Monday I retreated with grief & sorrow back again to Shrewsbury."[17]

Darwin must in many respects have been dreading reentry to Cambridge in October 1829, but he also had a plan. Whether by design or chance, he had found a much-needed new anchor. To the sequence of mentors there was now added someone who had been a teacher and friend of both his brother Erasmus and Fox: the remarkable Reverend John Stevens Henslow, fellow of St. John's College and professor of both mineralogy and botany. And with Henslow as his new guide, Darwin almost immediately changed again.

Darwin's connection with Henslow confirmed the pattern of his life so far. He worked well only when he had a mentor, not as a personal tutor but someone to bolster his confidence, someone to believe in him, and also someone safely to draw him out socially. With the right mentor, Darwin's underlying belief in himself could be brought out and channeled into productive paths; he could then engage fully with colleagues and teachers. Darwin's need for a mentor was so strong, and his weakness and loss of direction so pronounced when he lacked one, that one is forced to look back and wonder whether Darwin's cold dismissal of Jameson at Edinburgh and his apparent determination not to allow any credit for a positive influence on his geological training to go to Jameson may not mean that Darwin had sought out Jameson as a mentor and had been rebuffed. After Erasmus left Edinburgh, Darwin may have been drawn to Jameson as the one man whose interests were close to his: minerals, natural history, collections, science. But Jameson lacked the warmth of personality that would allow him to cultivate the very young Darwin, and Darwin in turn lacked the strength of personality to appeal to Jameson as an independent scholar. There is no way of knowing, but this interpretation would help explain the extraordinary dismissal of Jameson and Jameson's geology teaching in the *Autobiography*.

A parson and don, Henslow was only fifteen years older than Darwin and happily married. Presumably Erasmus had introduced Darwin to Henslow sometime early in 1828, before he left for London. Certainly Henslow had just the warm personal touch that Jameson and Grant had lacked and the lonely Darwin needed. After teaching for years in a Cambridge that did little to encourage the intellectual life except among the few superachievers, Henslow no doubt had a keen eye for students with a spark of originality. Henslow introduced Darwin properly to botany and also to more geology. And he introduced him to the society of scientifically inclined men at Cambridge. And Henslow loved to walk. By the end of his Cambridge career, Darwin had become Henslow's shadow, his alter ego: "the man who walks with Henslow."

Henslow also welcomed Darwin to his home, where he held evening soirées on Fridays in term time. Darwin did not start attending until his second year and was probably first taken to a soirée by Fox. He soon became a regular and steadily regained his "nerve," although he found Mrs. Henslow rather intimidating. At Henslow's house, over claret, Darwin met the great professors from Cambridge whose classrooms he had so far assiduously avoided—Adam Sedgwick the geologist, George Peacock the mathematician and astronomer. John Herschel from London, who was the dean of English science (and son of the astronomer royal William Herschel), was a regular attendee, as was Leonard Jenyns, who was Henslow's brother-in-law. Darwin's education blossomed in the traditionally Oxbridge way—not from lectures but from tutorials formal and informal, especially the latter. His health and sociability improved too. In January 1830 he could write to Fox, "It is quite curious, when thrown into contact with any set of men, how much they continue improving in ones good opinion, as one gets acquainted with them."[18]

As Darwin's new mentor, Henslow was everything that Grant could have been but was not: teacher, supportive friend, and intellectual guide. Under Henslow's tutelage, Darwin's natural history interests became more and more channeled. No longer a collector simply for the sake of collecting and the pleasure of the chase, Darwin became a more scientific collector, reveling in finding new things and especially when he discovered something new to science—for example, an extension of a species' range—or found new varieties. His hobby was turning into a science.

Darwin's competitive spirit never flagged. He was immensely proud when his discoveries made their way into a small notice in Stephens's *Illustrations of British Entomology*. For example, he wrote to Fox in July 1829, "You will see my name in Stephens' last number. I am glad of it if it is merely to spite Mr Jenyns."[19]

Intellectual life at Cambridge moved in a very different rhythm from that at Edinburgh, and with different preoccupations and fashions. Instead of lectures, the principal mode of instruction was through reading and tutorials. There was as yet no formal program of study in the natural sciences. All results depended on the final examinations, which inevitably required a great deal of last-minute cramming. Two routes were available. Students who intended a serious academic career were the "readers" trying for an honors degree. If they did well, the first two men in the lists were the First and Second Wranglers and had a chance of staying on with a college fellowship. Those who did not go for honors—by far the majority—were the "poll men." Darwin was in this group, but there was still a lot of work to do, a lot of catching up from time previously spent in nonacademic pursuits.

In his *Autobiography,* Darwin clearly wanted to give the impression that his academic labors were not arduous. He claimed: "In my second year I had to work for a month or two to pass the Little Go which I did easily," without mentioning that he had been advised to defer it for a year.[20] "Easily" was not a word that he used to describe the experience at the time, as his letters show. "I must read for my Little Go. Graham [Darwin's tutor] smiled & bowed so very civilly, when he told me that he was one of the six appointed to make the examination stricter, & that they were determined they would make it a very different thing from any previous examination that from all this, I am sure, it will be the very devil to pay amongst all idle men & Entomologists."[21]

Darwin was examined on "one of the four Gospels or the Acts of the Apostles in the original Greek, Paley's *A View of the Evidences of Christianity*, [and] one of the Greek and one of the Latin classics."[22] He finally took that examination in March 1830 (his seventh term), having studied intensely in the preceding weeks and completely neglected his correspondence with Fox. He wrote to apologize: "I am through my Little Go,!!! I am too much exalted to humble myself by apologizing for not having written before.—But I assure you before I went in & when my nerves were in a shattered & weak condition, your injured person often rose before my eyes & taunted me with my idleness."[23] Now Darwin had less than a year before he would take his finals, which students normally attempted in their twelfth term. For him, it would be two terms early.

Darwin had written of the final-year candidates of the preceding year: "The men are all reading at a most wonderful pace."[24] And: "The men are all in a dreadful plight, from fear & anxiety."[25] In January 1831, Darwin was in the same boat. "I worked with some earnestness for my final degree of BA, and brushed up my Classics together with a little Algebra and Euclid, which later gave me much pleasure as it did whilst at school."[26] In fact, he got Henslow to coach him in the always-troublesome question of algebra (he never could understand the binomial theorem).

For this final bachelor of arts examination, Darwin was examined in the classics (Homer and Virgil), in mathematics (Euclid and algebra), and Locke's *An Essay concerning Human Understanding*. He also had to "get up" two works by the great Cambridge teacher William Paley (d. 1805): his *A View of the Evidences of Christianity* (1794) and *The Principles of Moral and Political Philosophy* (1764). Paley's *Evidences* was one of the fundamental books of Anglican teaching and remained a required (or "set") book for Cambridge students for more than a century (the last time a final examination question was asked concerning *Evidences* at Christ's College was in 1920). In this work, Paley laid out the evidence for both the historical fact and the divinity of Jesus. His approach was that of a lawyer. Following Locke, whose philosophy greatly influenced him, Paley did not ask his readers to accept miracles but, rather, he based his case on the "eyewitnesses," the martyred saints of the early church. The whole approach is one of logical argument rather than appeal to faith. And that logic is what Darwin liked: "The logic of this book . . . gave me as much delight as did Euclid. . . . The careful study of these works, without attempting to learn any part by rote, was the only part of the Academical Course which, as I then felt and I still believe, was of the least use to me in the education of my mind."[27]

Paley's *Moral Philosophy* was a rather different proposition. Paley there took the very modern viewpoint that human institutions and human concepts such as right and wrong are in some way conditional rather than absolute. This was the utilitarian view, which measured the exigencies of human existence against what produced the greatest good for the greatest number, and utilitarian theses became very controversial in Darwin's time at Cambridge.

Overall, Darwin dismissed his Cambridge years: "During the three years which I spent at Cambridge my time was wasted, as far as the aca-demical studies were concerned, as completely as at Edinburgh and at school."[28] His tutor at Christ's College, John Graham, would have agreed. He had never been impressed by Darwin's diligence, judging him as "dawdling away his Cambridge days with his horse and his gun until thrown into a panic by the approaching examinations."[29]

Darwin took his final examinations in January 1831, having devoted all of the previous term to reading. On Guy Fawkes Day, he wrote to Fox, complaining, "I have so little time at present, & am so disgusted by reading that I have not the heart to write to any body. . . . I have not stuck an insect this term & scarcely opened a case."[30] On the 27th of November he wrote again: "I am reading very hard, & have spirits for nothing."[31] All the work paid off, however. He did quite well, coming in tenth of the 178 poll men who had not gone for honors. This was a more than respectable level of degree for an aspiring cleric who had made himself one of the country's leading amateur entomologists.

PART TWO

Savant

NINE

More Serious Things

It would be easy to get the impression from Darwin's *Autobiography* that his time at Cambridge was devoid of intellectual content, except for what was required for the pursuit and classification of a major collection of British beetles. He had no real commitment to developing his education in matters theological. He did not admit to having studied anything beyond the bare minimum necessary to pass his bachelor's degree examinations. The contrast between this and the careful, intense study at Edinburgh demonstrated in his surviving lecture notes from that previous life is striking, although the fact that he finished tenth on the list of poll men suggests that he had put in rather more effort than he pretended.

In many ways, the view that we have been given of Darwin's scientific life at Cambridge is so tame as to be thoroughly suspect. Cambridge may in many ways have been an undemanding and unexciting place in the 1820s and early 1830s, but it was not without controversy, and its dons did not fail to keep abreast of the latest developments in British and continental science. Nor was Cambridge immune to the political and religious disputes of the day.

It is not clear to what extent Darwin, who evidently shied away from all kinds of controversy, could keep himself apart from contemporary debates over political and social issues. At Edinburgh it would have been easier. Edinburgh was never so bound up with the party political scene as Cambridge; the Scottish capital had always viewed events in Westminster with a distanced eye. But to live anywhere in Britain in the

1820s without being aware of the political events of the time—from continued effects of the Corn Laws of 1815 on the economy and their ramifications for the growing army of the poor to the Poor Law reform of 1834 and the Reform Bill of 1831 in between, plus constant rumblings of revolution across the Channel—must have been well-nigh impossible. Even if he had tried to stay out of the fray, his mentor, Henslow, was one of the more prominent Whig radicals at Cambridge. Therefore, not everything about intellectual life at Cambridge can have been placid, with beetle collecting providing the only excitement.

In science there is always controversy. Henslow's Friday evening soirees were designed to provide a forum for like-minded men to discuss contemporary scientific issues. The Cuvier–St. Hilaire debate was one of the scholarly sensations of 1830; did it and similar events in the scientific world pass Cambridge by without causing a ripple of notice? Or was not Henslow's drawing room really a noisy, excited melting pot of ideas? Surely the great scientific ideas of the time were debated hot and loud *somewhere* over the tea and toast, or claret.

When Darwin wrote, "During the three years which I spent at Cambridge my time was wasted, as far as the academical studies were concerned," he definitely did not say that *everything* at Cambridge was a waste. The key to all this by-now-familiar self-deprecation and circumlocution lies in the term "academical studies." It was not possible to study formally for a degree in the sciences at Cambridge in Darwin's day. But a good deal of science was available to Darwin outside the course of formal academic study. As he admitted, "Public lectures on several branches were given in the university, attendance being quite voluntary." Henslow, for example, taught lecture courses in both mineralogy and botany, and Darwin's name shows up on the rolls for Henslow's course in botany for three years in a row: 1829, 1830, and 1831. Although he always claimed a distaste for lectures as a teaching genre, Darwin "liked [Henslow's lectures] *very* much for their extreme clearness, and the admirable *illustrations.*" He also added, carefully, "but I did not study botany." Darwin particularly enjoyed the fieldwork: "Henslow used to take his pupils, including several older members of the university, on field excursion, on foot, or in coaches to distant places or in a barge down the river, and lectured on the rare plants or animals which were observed. These excursions were delightful."[1]

If taking Henslow's course three times was not to "study botany," it is hard to know what (except in the sense of a formal degree) *studying botany* might have been. In Henslow's botany course he made himself so obnoxious trying to be the "teacher's pet" that other students were ready to strangle him. Whether it was something as mundane as arranging the chairs in the lecture room or vying to ask the most penetrating questions, Darwin seems always to have been promoting himself in Henslow's eyes. (Darwin's principal competitor for Henslow's favor was his collecting rival, "Beetles" Babington, who might be considered to have won this competition, eventually becoming Henslow's successor as professor of botany.)

Insect collecting was not without its intellectual and theoretical side. When Darwin was a student, a great deal of interest was generated by MacLeay's *Horae Entomologicae* (previously mentioned), in which the author attempted to derive a whole "new system of nature" out of a case study of the classification of the scarab beetle family.[2] MacLeay was a graduate of Trinity College, Cambridge, from which he had entered public service, all the while developing as a leading entomologist. His ideas had rapidly become popular, and accounts of them appeared in the work of men like William Swainson, who wrote extensively on natural history in the 1830s. It was attractive because it attempted to introduce an intellectual rigor to what had previously been a highly subjective "art."

MacLeay's book dealt with the key issue in philosophical zoology and botany: finding a "true" and "natural" basis for a scheme of classification. Creating reliable systems of classification depended on sorting out two kinds of similarity: a "true" relationship or *affinity,* and a false one or *analogy.* This was an ancient problem. Greeks like Aristotle, for example, had to think hard about whether to classify the whales with fish or with land mammals (as whales have lungs and suckle their young, they belong with mammals—but they look like fish). Then there are the wings of bats and birds. In fact, the problem was almost impossible without knowing a rational basis for the causes of such similarities and differences among species. It was precisely such a cause that Darwin would eventually enunciate. Meanwhile, MacLeay (and Darwin at that time) thought that all species had been created individually by God, which left the scientist in the position of someone trying to put together a giant jig-saw puzzle.

The Quinary System was intended not just as a formal extension of the principles of classification but as the basis for a whole "system of nature," affording a key to finding "that uniformity of plan, upon which every object in nature was originally created."[3] It was based, somewhat mystically, on the number five. MacLeay thought that, at all hierarchical levels, taxonomic relationships could be portrayed in patterns of five contiguous "circles of affinity." In such circles, as in tracing longitude on a globe, if one went far enough round the circle, one came back to the beginning. At the lowest level, species could be grouped first into subgenera, geographically constrained. The circles of subgenera could then be arranged into genera, the genera into families, and so on. Each "circle of affinity" (that is, a group of truly related organisms) touched two others, and between them were key "osculant" groups, having some characters of each circle. In addition to MacLeay's osculant groups, Swainson identified the condition of inosculation. The inosculant groups were those taxa within the circles that touched the osculating groups. As various people (including Darwin) used the terms, they tended to confuse them.

The influence of MacLeay's Quinary System was seen everywhere in British natural history from 1820 to about 1850. For example, in the first volume of his *Illustrations of British Entomology*, James Stephens used a quinarian reference with respect to a large group of insects: "These appear to be connected together in affinity . . . the series 'returns into itself.' "[4] Although there is no direct reference to MacLeay in Darwin's letters and notes from his years at Cambridge, Darwin was familiar with quinarian concepts. He wrote whimsically to Henslow from HMS *Beagle* in 1832 about a bird he had seen: "a happy mixture of a lark pigeon & snipe.— Mr Mac Leay himself never imagined such an inosculating creature."[5]

A parallel subject with potential for controversy was one with which Henslow was deeply concerned: the nature of the species itself. He

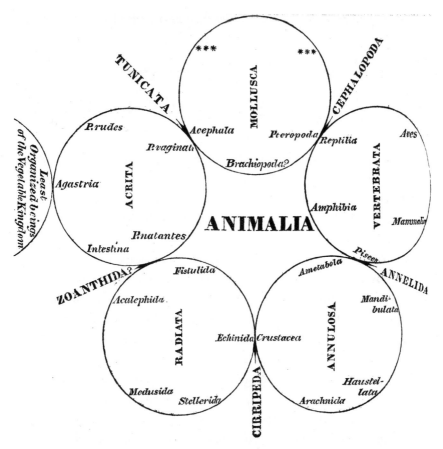

2. William Sharp MacLeay, in his Quinary System, arranged all groups of
animals in systems of five. In this overall view, the "osculating" groups are
indicated with an arrow (for example, Cephalopoda, which link vertebrates and
mollusks); "inosculant" groups are those within the circle that meet the
"osculants." *Horae Entomologicae* (London: Bagster, 1819), 1:318.
(Courtesy of the Ewell Sale Stewart Library, Academy of Natural
Sciences of Philadelphia.)

made it the subject of both his lectures and his field excursions. In biology, the concept that animals and plants could be characterized as belonging to discrete, formal entities called species, and that species were
the building blocks of systems of classification (the basic ordering of
knowledge), had been a major advance of eighteenth-century science.

What the ancients had called "kinds" had been accorded a position of highest scientific priority. But, ever since Linnaeus's seminal work in taxonomy, there had been a number of deep divisions among scholars about the nature of species. Lamarck, for example, thought that the sum total of all species, when discovered, would eventually turn out to be a collection not of discrete entities, but of entities that graded imperceptibly into each other. For him, then, the concept of species ("kinds") in animals and plants was a convenience, not a truth. Those devoted to avowing the literal truth of the Bible assumed on the contrary that, because God had made the species, he made each one uniquely different, their characters fixed and unchanging.

Empirical evidence showed that, as Linnaeus had also pointed out, species are, in fact, not invariant; their constituent members differ significantly among themselves. Furthermore, a growing body of evidence had shown that these variations usually represented adaptation to local conditions in the environment. If species were highly variable, particularly in those characteristics that were used to classify them, practical biology was in something of a difficulty. The same problem would apply in geology where, following William Smith and others, characteristic species of fossils were used uniquely to identify discrete strata.

Amateur natural historians (many of them no doubt hoping thereby to achieve certain fame) tended to describe a lot of new species on the basis of very small differences. They named dozens of species where, according to cooler heads, there were but a few, or even one.[6]

All this made it crucial for any serious taxonomist to learn the extent, and context, of intraspecific variation. Just at the time when Darwin was at Cambridge, Henslow was one of the leaders among those trying to show the falsity of the extreme "splitter" position and conducted experiments to test the situation. Plants are better experimental organisms for this work because seeds can be collected and then grown in a range of artificial conditions. Henslow, in collecting plants from all over Britain, had already seen that many species tended naturally to vary extensively. The primrose/cowslip group of "species" was a special interest of his. He referred to a study in which seed from a "red cowslip" had been grown in a highly manured bed, producing "a primrose, a cowslip, oxlips of the usual and other colors, a black polyanthus, a hose-in-hose cowslip, and a natural primrose bearing its flower on a polyanthus stalk." On one of his botanical excursions around Cambridge, Henslow then found a patch of highly variable "cowslips" that he bred at home in his garden with similar results, concluding that all these "species" were the variable manifestations of a single type.[7]

Another case that he investigated at that time concerned the wild-flowers known as pimpernels in the genus *Anagalis*. The distinguished botanist William Hooker had distinguished two species—*A. arvensis* (red) and *A. cerulean* (blue)—on the basis of flower color and petal shape. Henslow "got some seeds of blue *Anagalis cerulean* and raised a dozen plants nine of which have blue flowers, and three have red." From a white variety he raised "one which flowered red, and the other six white . . . and having a bright pink eye." Henslow concluded that the supposed species were only varieties. But he used his experiments to articulate as a "law in botany . . . that if a change takes place in one of the organs of a plant, a simultaneous change may be expected in some or all of the organs . . . for instance, when we see the leaf of the cowslip differing from that of a primrose, we need not be surprised to find that the calyx and corolla should differ also."[8] Many years later, Darwin did some crucial work on these same forms, and in *On the Origin of Species* he devoted a whole chapter to "Laws of Variation" like the one proposed by Henslow.

Henslow's attention to the subject of the variability of species neatly ex-emplifies the difference between natural history and natural philosophy. As a natural historian, Henslow was concerned with establishing the facts of variation within species and with the practicalities of plant tax-onomy. His religious convictions allowed him to avoid the logical con-sequence of variation, namely, that it might be a mechanism in the formation of new species. It was the kind of intellectual compromise into which English academics were increasingly being forced.

The most celebrated example of this "English Compromise" con-cerned geology, which had been maturing fast as a science since 1800, producing controversies that Darwin could not have avoided.[9] The dons at Oxford and Cambridge necessarily stuck to their Christian (or surreptitiously deist) view of the authenticity of the biblical record. Orthodox opinion held that regardless of the age of the earth, it had been created by God; all life had been created by God; there had been a Noachian Flood that caused great changes in landscapes and animals and plants; all species were God's creation; humans were God's last work. All new geological information had to be fitted—sometimes easily, sometimes with difficulty—into this view.

In Darwin's student days, there were two great professors of geology in England: William Buckland at Oxford and Adam Sedgwick at Cambridge. Both were clerics. Buckland was a great teacher and writer; his books, such as *Reliquiae Diluvianeae* (Evidences of the Flood, 1823), brought geology to a wide audience outside of academe because they started from a biblical view of the Flood. Buckland accepted that the earth was very old, which meant that there was a gap in the account of Creation in Genesis. Between the initial Creation and modern times there had been a whole series of catastrophes, shaping the earth and life upon it, of which Noah's Flood had been the last. This put Buckland in opposition to the views of James Hutton that the earth had only ever been shaped by the sorts of processes that we see in action today.

Adam Sedgwick graduated from Cambridge in 1808 and took holy orders. He did not train as a geologist; in fact, he knew virtually nothing of the subject when he was prevailed upon to take up the Woodwardian Chair. Such was the way in which Cambridge could award honors to its more promising sons. Rather than accept a sinecure, Sedgwick taught himself and, by investing a great deal of time in field studies, rapidly became a force in British geology. By 1829, he was president of the Geological Society of London. An attractive, slenderly built man with an engaging personality, Sedgwick commanded respect within the university and without, even though his views were often very liberal.

In 1830, the virtual monopoly over the field in Britain that Jameson, Buckland, and Sedgwick enjoyed, despite their different views of how geological science fit into a religious framework, was challenged by an outsider. Charles Lyell came from a wealthy Scottish family and had studied with Buckland at Oxford before going to London to practice law. In his mid-twenties, he decided to follow his heart and take up geology seriously. Less than a decade later, he published the first volume of his *Principles of Geology*, a work that was intended to bring the whole field under one rational system. It drew its authority from his extensive fieldwork in Europe and his exhaustive command of the literature in English, French, German, and Italian.

In his religious views, Lyell was essentially a deist, holding the position that God had originally created the world and life on it, and then had allowed nature to operate according to its own (God-given) natural laws, rather than constantly intervening to direct and shape the course of all history. He stated, therefore, that studying the Bible was of no use whatsoever in trying to understand the geological history of the earth.

Lyell's central principle was uniformitarianism: the earth had been shaped gradually by the same forces that we see acting today, and at the same rates and intensities, rather than by a few dramatic, all-embracing convulsions (catastrophism).

Lyell cut through the debate between the Neptunists and Plutonists and took as his starting point the second of James Hutton's theses, seeing earth history as essentially cyclic, rather than, as Judeo-Christian dogma required, unidirectional and finite. The earth is shaped by a steady process of elevation and subsidence due to earthquakes, and erosion and deposition of strata.[10]

Apart from catastrophists and Flood theorists, one of Lyell's chief targets was Lamarck (whose works had been available in English for a decade), and all those who saw in the fossil record a progression ("successive development") of organic life on earth. Ever since Aristotle, a mainstay of natural philosophy had been the concept that all nature was constituted as a Great Chain of Being, a hierarchical arrangement of life from highest (humans being the highest and next to the angels) to lowest (worms and insects).[11] (Most recently, Cuvier and Lamarck had produced scientific classifications that similarly arranged animals hierarchically, from worms to fish to reptiles to birds to mammals and finally to humans.) Then, over the first decades of the century, the fossil record had started to show that the different groups of organisms had first appeared on earth in this same sequence. This view was endorsed by Jameson in his influential edition of Cuvier's *Essay on the Theory of the Earth*. A related idea, growing in popularity around 1830, was that the succession of fossils, of both animals and plants, demonstrated a shift in climate from hot (older—the Coal Measures, for example) to colder (the present).

 Obviously, one reason for attacking "progression" or "advancement" was that, like variation within species, it led logically to thoughts of transmutation of species. Lamarck explicitly used progression to argue for his ideas on transmutation. The anonymous articles on evolution that had appeared in the *Edinburgh New Philosophical Journal* (Jameson) had espoused those ideas. In the second volume of his *Principles of Geology*, Lyell devoted much space to dismantling Lamarck's views. In so doing, he helped publicize them!

Lyell's rejection of a unidirectional progressive earth history led him to an odd conclusion concerning extinction. He thought that there was evidence for climate change in the history of the earth, but it had been cyclic. As there had been no linear progression, "then might those genera of animals return, of which the memorials are preserved in the ancient rocks of our continents. The huge iguanodon might reappear in the woods, and the ichthyosaur in the sea, while the pterodactyl might flit again through umbrageous groves of tree ferns."[12] The drama of this statement was increased by the fact that many of these fabulous creatures had been revealed in the fossil record only in the decade or so before Lyell wrote: the first complete *Ichthyosaurus* in 1811; the dinosaurs *Megalosaurus* and *Iguanodon* in 1824 and 1825, respectively.

At Cambridge, where instruction was centered on theology and the classics, there were lots of reasons why the cleric-dons would find Lyell's ideas positively dangerous. He was attacked principally for his rejection of any value in a biblical approach to geology; for his conclusion that the earth was far older than any biblical scholar imagined;

and (in his rejection of the role of catastrophes) for his dismissal of the supposed Noachian Flood.

Darwin, clinging somewhat tenuously to sufficient Christian faith and belief in order to make a career as a country parson feasible, evidently decided to go along with the majority and found the English Compromise acceptable and even, in many respects, rational. However, once he started to attend Henslow's soirees, the swirl of ideas was brought closer home.

Adam Sedgwick, one of the more politically liberal dons and a close friend of Henslow, was a regular participant in the Friday evening discussions. He believed firmly in the historical fact of Noah's Flood. Although he had never thought that the Flood explained all of geology, as the Mosaic literalists did, he saw direct evidence for the biblical Flood in the vast deposits of sand and gravel that cover much of western Europe. For more than a hundred years, scholars had argued that the Flood described in the Bible, if it had occurred at all, would not have covered the entire earth but must have been local to the Middle East. There could not have been enough water in forty days and nights of rain, or even four hundred days and nights of rain, to create a worldwide flood. Even if the earth had broken open to release the "fountains of the deep," that, in turn, required explaining where such water had come from and where it went to. (Many earlier scholars, from Descartes onward, had posited that the earth's crust covered a layer of subterranean water.) There was, therefore, already tension among geologists, believers or skeptics, about the Flood, but the extensive deposits of sand and gravel clearly denoted that some kind of water-driven process had acted on a global scale.

In 1825, Sedgwick's view of the Flood had been entirely conventional, even reactionary. "The sacred record tells us—that a few thousand years ago 'the fountains of the great deep' were broken up—and that the earth's surface was submerged by the water of a general deluge; and the investigations of geology prove that the accumulations of alluvial matter . . . were preceded by a great catastrophe which has left traces of its operation in the diluvial detritus which is spread out over all the strata of the world. . . . Between these conclusions, derived from sources entirely independent of each other, there is, therefore, a general coincidence which is impossible to overlook, and the importance of which it would be most unreasonable to deny. The coincidence has not been assumed hypothetically but has been proved legitimately, by an immense number of direct observations conducted with indefatigable labor, and all tending to the establishment of the same general truth."[13]

Nonetheless, in his continued assiduous application to fieldwork on British geology, Sedgwick began to confirm what had long been noticed. The great deposits of "diluvial" sand and gravel were not uniform but had evidently been deposited in discrete layers of different types. This could only mean that whatever had caused them had not been a single, short-lived event; the process or processes that had laid them down must have acted intermittently over thousands of years.

In his address to the Geological Society on February 18, 1831 (shortly after Darwin took his finals), Sedgwick announced a complete change of mind. He still believed that there had been a biblical Flood. "Our errors were, however, natural, and of the same kind which led many excellent observers of a former century to refer all the secondary formations of geology to the Noachian deluge. Having been myself a believer, and, to the best of my power, a propagator of what I now regard as a philosophic heresy, and having more than once been quoted for opinions I do not now maintain, I think it right, as one of my last acts before I quit this Chair, thus publicly to read my recantation. We ought, indeed, to have paused before we first adopted the diluvial theory, and referred all our old superficial gravel to the action of the Mosaic flood."[14]

The news of Sedgwick's volte-face must have caused shock waves in Cambridge society. And Darwin had the great good fortune to have been at Cambridge and a peripheral member of one of its inner circles at that time. One can only imagine the effect on the participants at Henslow's soirees, already reeling from Lyell's assaults, as Sedgwick patiently explained his thoughts and his reasons for still managing to maintain a belief in some kind of biblical Flood.[15]

Cambridge had turned out to be no less contentious a place than Edinburgh. Sedgwick himself summarized the issues that were most troubling to dons and students alike in the years when Darwin was a student at the university. In a famous book, *A Discourse of the Studies of the University of Cambridge,* he confirmed how important was one of these issues—none other than transmutation of species.

Sedgwick's little book started out as a two-hour sermon he gave in Trinity College chapel on December 17, 1832, on the occasion of the annual commemoration of the college's founders. He was soon prevailed upon to publish, and it eventually went into five editions. In one sense, Sedgwick's thesis was quite simple: it was the old compromise. For students at Cambridge, "intellectual studies—properly interpreted—were not at all inconsistent with their religious beliefs. On the contrary, science and philosophy were at all points . . . moral and religious habits of thought."[16] (A very great deal, of course, depended on the phrase "properly interpreted.")

Sedgwick's message was also very complicated. *Discourse* was a donnish revolution of its own. In attacking a range of popular "heresies," from agnosticism to biblical literalism, that threatened the balancing act of compromise, Sedgwick ensured a wide readership by building his argument upon a frontal attack on a central pillar of the Cambridge curriculum: William Paley's *The Principles of Moral and Political Philosophy,* one of the books on which Darwin had been examined. Sedgwick was affronted by Paley's view that right and wrong were not fundamental truths embedded in us by God but merely contingent—determined by what achieved the greatest good in particular circumstances. His attack on Paley, even if accompanied by admiration for his other works, was almost scandalous. The ensuing arguments over the role, and the very meaning, of utilitarianism drew in John Stuart Mill and others. Sedgwick had opened a hornet's nest.

Before launching into his discussion of moral philosophy, however, Sedgwick focused on another subject—one almost as close to his heart as the fundamental bases of right and wrong. He set out his views on current geological heresies and the new studies in geology that tested the old compromise by creating all sorts of problems for believers in the literal truth of the Bible.[17] Prime among these was Lyell's argument against progression ("successive development") in the fossil record. Contra Lyell, Sedgwick saw the fossil record as linear and progressive. In arguing this point, he wrote in language that seems to modern eyes both to leave him open to the charge that he was guilty of the same reasoning from blind faith that had led to his incorrect view of the Flood, and at the same to time to flirt with the vocabulary of evolution.

Sedgwick maintained the orthodox Christian interpretation that there had been "progression" and that it was due to God's constant intervention. "Geology tells us that God has not created the world and left it to itself, remaining ever a quiescent spectator of his own work: for it puts before our eyes the certain proofs, that during successive periods there have been, not only great changes in the external conditions of the earth, but corresponding changes in organic life. . . . It shews intelligent power—at many times contributing a change of mechanism adapted to a change of external conditions; and this affords a proof, peculiarly its own, that the great first cause continues a provident and active intelligence. . . . If our planetary system was gradually evolved from a primeval condition of matter . . . every change . . . has been but a manifestation of the Godhead."[18]

In the fifth edition of his *Discourse,* Sedgwick took on Lyell even more explicitly: "There are traces among the old deposits of the earth of an organic progression among the successive forms of life . . . absence of mammalian in the older, and their very rare appearance in the newer Secondary groups, . . . lastly, the recent appearance of man on the surface of the earth. . . . This historical development of the forms and functions of organic life during successive epochs seems to mark a gradual evolution of creative power, manifested by a gradual ascent towards a higher type of being."[19] But, even as late as the 1850s, Charles Lyell continued to deny that the fossil record showed "progress," arguing that, for example, higher tetrapods were no more highly developed than complex Devonian fish.[20]

In addition to taking on Lyell, Sedgwick also attacked Lamarck and, in so doing, demonstrated that transmutation had become a threatening subject of active discussion at Cambridge. Grant had brought Edinburgh Lamarckism to London, and the influence of Paris on British intellectuals was growing. Like Lyell, the core of Sedgwick's opposition to Lamarck was the latter's conclusion that "progression" necessarily implied "transmutation of species." Unlike Lyell, Sedgwick had opened the door to the fact of "progression"; he now had to close off the logical consequences.

What he wrote in contradiction of Lamarck was noticeably not based on fact but mere assertion: "The elevation of the fauna of successive periods was not made by transmutation, but by creative additions; and it is by watching these additions that we get some insight into Nature's true historical progress." And this was especially true of human origins: "We therefore believe that man, with all his powers and appetencies, his marvelous structure and his fitness for the world around him, was called into being within a few thousand years of the days in which we live—*not by a transmutation of species (a theory no better than a phrensied dream)* but by a provident contriving power" (emphasis added).[21]

The sum of all this was to demonstrate that issues like the nature of species, the Flood, progressive development, the age and history of the earth, and transmutation were all topics of active discussion in Cambridge in 1831. We may wonder, therefore, just how, in the Reverend John Henslow's living room, after Sedgwick's apostasy and with transmutation lurking unbidden in the wings, colleagues like the Reverend Adam Sedgwick, the Reverend George Peacock, the Reverend Leonard Jenyns, the Reverend William Whewell, and would-be-reverends like Darwin dealt with these burning issues. Or were they more concerned with the social and political events—the poor, the Reform Bill, and university politics—than the deep philosophical questions of science? And how was the budding intellectual, Charles Darwin, actually spending his time?

TEN

Reading Science

Because he had matriculated late, even though Darwin passed his examinations on January 22, 1831, he was not "admitted to the degree" until April 29 (and his name counted among those who graduated in 1832). Forced to remain in residence for two more terms, Darwin found that he had come to a crossroads in his life. Choices had to be made. He could have remained an idle sporting man, perhaps devoting even more time to building and classifying his insect collections. He could have become little more than a playboy. He might have spent this enforced residence in Cambridge buried deep in theological matters in order to tackle the next hurdle on the route to becoming a parson. Instead, under Henslow's guidance, Darwin began to delve deeper into the more scholarly aspects of natural science and particularly the connections between natural history and natural philosophy. Whatever he had been before this—and it is hard to believe that he was not at least a serious and eclectic reader—at this point Darwin's intellectual development changed in ways that directly affected his later career as a scientist.

The remarkable part of this is not that he chose this course but that he followed it with such vigor and intensity. He chose this moment, according to the *Autobiography,* to abandon all *appearances* of fecklessness. From this point on, Darwin allowed himself to be fully revealed as someone who was deadly serious about himself. Even when he was casual, it was deliberate.

In his *Journal,* Darwin noted: "During these months lived much with Prof. Henslow, often dining with him, & walking with. Became slightly acquainted with several of the learned men in Cambridge.—which much quickened the little zeal, which dinner parties & hunting had not destroyed. . . . Henslow persuaded me to think of Geology & introduced me to Sedgwick."[1] These notes were probably written years later and, as usual, are inconsistent. He had been attending Henslow's soirees since the autumn of 1829. He surely met Sedgwick and others there before 1831, and he attended at least some of Sedgwick's geology course. That was when he took Henslow's botany for the third and last time.

Under Henslow's guidance, Darwin first launched himself (characteristically) into further reading. He did not study the classics, theology, or the ever-problematic mathematics, but stuck to natural science, studying three works in particular (at least, we know about these three because Darwin singled them out in the *Autobiography*—there must have been many more). In February he read John Herschel's recently published *Preliminary Discourse on the Study of Natural Philosophy* for its brilliant exposition of the scientific method. That same month, he also read Alexander von Humboldt's *Personal Narrative of a Journey to the Equinoctial Regions of the New Continent,* which he had read before and continued to read and reread for its evocative descriptions of the tropics and a holistic view of science that gathered geology, geography, botany, zoology, and climatology within a single compass. Finally, Darwin singled out for mention *Natural Theology; or, Evidences of the Existence and Attributes of the Deity* by William Paley, the third book in his trilogy (with *Morals* and *Evidences,* which Darwin had already devoured and approved). Darwin quickly came to appreciate its logical development of "the argument from design," which saw in the works of nature evidence for the existence and purposes of God. "I do not think I hardly ever admired a book more than Paley's 'Natural Theology.' I could almost formerly have said it by heart."[2]

When Sedgwick, in his *Discourse,* had criticized William Paley for his utilitarian approach to morals, he also had been at great pains to praise, and even steal extensively from, his *Natural Theology*. This was Paley's last book, and its aims were hardly modest. Paley attempted to use an overwhelming mass of data and ideas from the natural sciences to "prove" that God is the only possible explanation of the cause of life and to reveal the very nature of God. *Natural theology* was a term that had been in use for more than a century. Paley's book was an up-to-date statement of the same arguments that had been put forward by John Ray in *The Wisdom of God Manifested in the Works of His Creation* (1691) and William Derham in *Physico-Theology; or, A Demonstration of the Being and Attributes of God, from His Works of Creation* (1716). Paley rehearsed and extended John Ray's arguments, armed with a century's worth of new knowledge about the structure and physiology of living organisms. His *Natural Theology* could be (and was) used as a textbook of biology. The natural history exhibits of the Ashmolean Museum at Oxford were once arranged in the order of Paley's chapters.

The essence of *Natural Theology* is nothing more or less than the same "argument from design" that has attracted philosophers from ancient times to today. As one contemporary reviewer wrote, "On the subject of Natural Theology, no one looks for originality, and no one pretends to discover it. Its great virtue is its great simplicity."[3] The central premise of the book is a simple syllogism: complex machines (a watch, for example) require an intelligent designer; living organisms are complex mechanisms; therefore, living organisms have an intelligent designer. Something like Paley's central metaphor—a watch found lying on the ground—could not have arisen by chance or have created itself; therefore, neither could living organisms.

The first conclusion of *Natural Theology* was therefore that the sheer complexity and diversity of nature requires the "provident and active intelligence" to which Sedgwick referred. God must exist and must have been that architect. The second conclusion amplified the first: God's actions and "attributes" are demonstrated in the matter of "purpose." The pillars of Christian faith are that God was the Creator, that his Creation has a purpose, and that all life is directed toward a final End. Christ was his intermediary, sent to show sinful man the path to redemption. Paley documented the fact that the material objects of living nature—animals and plants, cells, organs, functions, behaviors, from those of the simplest organism to humans—all have one consistent feature. They show an apparently perfect match of form and function. He then argued that these adaptations—the eyes for seeing, wings for flying, fins or webbed feet for swimming, the beak of the hummingbird, and so on— are manifestations of the purposefulness of God.[4]

Natural theology is the foundation of the modern concept of "intelligent design," espoused by antievolutionists and creationists. A favorite example of "design" is usually taken to be the eye, which Paley saw as a nice analogue of the telescope (it is even more like a modern digital camera). Darwin would later devote many pages to this tricky subject, and the challenge of showing what a partially evolved eye might have looked like or been good for, in *On the Origin of Species*. Today, supposedly irreducibly complex features of organisms (that, by inference, only God might have created) are looked for in the intricacies of biochemical pathways and microscopic cell structure, but with no possibility of finding a "proof."[5]

Paley wrote so clearly and thoroughly, and he treated the subject so rigorously and scientifically, that he seemed to many to have convincingly defeated the arguments of the atheists. Indeed, Darwin, the intended clergyman, read Paley with enthusiasm, finding his arguments persuasive.

Paley's book is also a curious political statement. In order to describe the "attributes" of God, Paley had to explain away the less-than-perfect nature of the real world. It was all well and good for Paley, in full utilitarian mode, to argue that "teeth were made for feeding, not for toothache," but teeth were also made for biting, for tearing flesh. The predator was happy, the prey wretched. In the human world, he had to account for the miseries of war, pestilence, famine. The violence of nature and man's inhumanity to man had somehow to be accounted for as God's beneficent will. Paley had to argue that everything happened for the best in God's world.

In one of his more important arguments, Paley harnessed the new ideas of Thomas Robert Malthus on the natural and voluntary control of human population numbers.[6] Cleverly, Paley thereby shifted all the "bad" in life to the operation of second causes. Phenomena in nature that appear bad to us result from two lawful, God-given principles: the shifting "oeconomy of nature" and superfecundity (the capacity to produce more offspring than can possibly survive). Together they produce a struggle for survival. This is the first mention of the phenomenon that would be central to Darwin's theory of natural selection.

Paley had first opposed Malthus's ideas on population because he imagined that nothing could be better for happiness and success than an abundance of population; for Paley, the more the merrier. Malthus's *Essay* logically destroyed that argument. Malthus showed very convincingly that population increase was natural (populations tend to grow at geometric rates), but food and other resources grew more slowly (arithmetically). Population size was kept in check by war, pestilence, and famine as well as (Malthus thought to a lesser extent) by preventative checks like changed behavior. In this last case, Malthus had in mind delay of marriage or increased intervals between children, not, as later authors have thought, birth control (a vice and sin to Malthus).

Following Malthus, Paley explained that human and animal hardship was (to put it nicely) part of the necessary God-given order of things. "There are strong intelligible reasons, why there should exist in human society great disparity of wealth and station."[7] This argument was in direct opposition to the revolutionary trends emerging from France and from English writers like Thomas Paine, whose ideas on self-improvement and progress included the possibility of breaking down the traditional class order.

Progress and self-improvement were anathema to the established order and therefore to the English oligarchy. Paley was here making an argument for the "natural" rightness of the status quo ante, from the church to the House of Lords, and indeed for its natural (God-given) origins. All this was possible because Paley was quite sure that it was just as possible for a poor person to be happy and fulfilled as a rich one. Wealth and station had nothing to do with true human contentment which, again, grew from matters spiritual that God ordained.

There was more. Paley's *Natural Theology* is an argument aimed squarely at the particular kind of freethinking heresy that was promulgated by Darwin's grandfather, Erasmus, and by Buffon before him. No doubt Darwin read those sections with particular interest. Paley took aim at the apparently absurd, atheistic idea that "[organs and organisms in] every organized body which we see, are only so many combinations out of the possible varieties and combinations of being, which the lapse of infinite ages has brought into existence; that the present world is a relict of that variety; millions of other bodily forms and other species having perished, being by defect of their constitution incapable of preservation, or of continuance by generation."[8]

This last was, of course, Erasmus Darwin's "heresy," and the substance of the last sentence reappeared in exactly the same form—selection—in Charles Darwin's theory. But the only argument Paley could muster against it is that we cannot see this in nature: "There is no foundation whatever for this conjecture in any thing which we observe . . . no such energy operates . . . pushing into existence new varieties of being." Then he descended to sarcasm: "A countless variety of animals might have existed, which do not exist. Upon the supposition here stated, we should see unicorns and mermaids, sylphs and centaurs. . . . Or, if it be alleged that these may transgress the limits of possible life and propagation, we might, at least, have nations of human beings without nails upon their fingers, with more or fewer fingers or toes than ten."[9]

One problem with the attempted demolition of a theory is that, in order to show its worthlessness, one has to discuss it in detail. And in so doing, as in the case of Lyell, Lamarck, and "successive development," one draws attention to the theory—so that readers may take it far more seriously than if the theory had been ignored. And in taking the criticism seriously, one can turn the theory around and build a better one. That is exactly what Darwin eventually did with Paley's views of the stabilizing effect of the "oeconomy of nature"—making it an agent for change.

One wonders what Darwin's reaction was in 1831 to all this high-handed dismissiveness. He had no doubt heard all the arguments before, but here they were expressed in a book whose logic he professed to have enjoyed and welcomed. When his grandfather's ideas were savaged, did he happily agree and close the book, or did he secretly, perhaps unconsciously, say to himself, "In that case, I can see what sorts of argument would be needed to make Grandfather's point. A lot of what Paley says is obviously rather contrived—this makes me wonder . . ." Certainly he would have said to himself, "This is not fair. *Zoonomia* cannot be rejected so simply, out of hand, with sarcasm, even if it is wrong." If so, as in the Bible, he kept all these things in his heart (and possibly in his notebooks, although none of his notes on Paley seem to have survived).

Humboldt's *Personal Narrative* (published in installments between 1815 and 1825; complete English translation, 1829) was a completely different work from Paley's lawyerly argument. It was one of the most successful popular travel books of all time, surpassed only by Darwin's own *Voyage of the Beagle.* It is full of the glories of the tropics and the potential of biological science. Reading Humboldt opened Darwin's eyes to the world of natural history outside of the British Isles and directed his intellect to the vastness of biological diversity worldwide, which previously he had known only from museums (Jameson's, of course).

Humboldt's *Personal Narrative* might have been exciting travel writing to others, but to Darwin it represented the meeting of two different approaches. Humboldt was both a serious scholar and someone whom the amateur—the country parson, perhaps—could emulate. Humboldt was Henslow writ large, another model to be followed by a person of the right ambition.

The actual style of Humboldt's *Narrative* seems almost too eclectic for modern eyes. It seems far more dated than Darwin's book of two decades later. At its best, it mixed rambling observation and descriptions with both a gentle romanticism and a harder scientific edge. To contemporary readers, the book, with its descriptions of South America, its adventures and dangers, was entrancing. It could hardly have been a better model for Darwin's own *Voyage of the Beagle.* But there was more to Humboldt in Darwin's eyes than novelty and a kind of international superstar glamour. Darwin first met Humboldt, so to speak, at Edinburgh—as a scientist, geologist, geographer, and climatologist— in the pages of the *Edinburgh Philosophical Journal* and *Edinburgh Journal of Science*. A range of Humboldt's work appeared in these journals for the English-speaking audience, including his ideas on the geology of mountains.

Humboldt had been a student of Werner and was centrally involved in discussions of what we now call orogeny—the causes of the creation of mountains and particularly mountain chains. In the 1820s, this was one of the most hotly contested subjects, with rival theories ranging from the extreme Wernerian view that present mountain chains were simply the remaining parts of the original land masses to the extreme Vulcanist view that they were thrown up by volcanic activity and earthquakes and that any ancient mountains had long since been eroded. Of the contending views, once the Neptunist view declined, about whether mountains were raised up by volcanoes or by earthquakes, Humboldt, who studied the worldwide geography of volcanoes, tended to the latter view.

Darwin's interest in Humboldt at Edinburgh was, of course, in large part due to his teacher Jameson. Humboldt's great talent was in synthesis—in bringing together highly original work in the otherwise diverse subjects of geology, geography, climate, zoology, and botany (both taxonomy and physiology). "I shall try to find out how the forces of nature interact upon one another and how the geographic environment influences plant and animal life. In other words, I must find out about the unity of nature."[10] This was exemplified in his studies of the relations between plant distribution (and habit) and altitude on mountains. He developed the concept of isothermal lines—similar to the isobars of atmospheric pressure familiar in any modern weather map. The distribution of plants was shown to be not the simple two-dimensional function of geography, but a three-dimensional issue. Darwin had read and digested these works of Humboldt when in Edinburgh. When later assembling a library of scientific papers for the voyage, he specifically asked his sister Susan to send him from his bedroom Humboldt's paper on isothermal lines.

Humboldt was seriously interested in everything. One could say exactly the same about Darwin. Moreover, Humboldt was more than just a successful scientist; he was an international figure, a hero of science and exploration. Did Darwin have an ambition to make a similar place for himself in science, back in the spring of 1831? I suggest that it was precisely during this period that he discovered that ambition, again through models and mentors, but at that time he was not so ambitious as to aim as high as Humboldt's "unity of nature." Darwin had the yearning for greatness but not yet the means to achieve it. But, under the immediate influence of Humboldt, and reflecting his long interest in travel and exploration, he started to plan a tropical expedition of his own.

If Alexander von Humboldt inspired Darwin from afar, John Frederick William Herschel was a Cambridge man and someone Darwin actually knew. His *Preliminary Discourse on the Study of Natural Philosophy* was a popular, rather than purely scholarly, book. It was widely influential as an exposition of the nature of the scientific method and the philosophical bases of scientific investigation. It distilled exactly what Darwin always meant when he referred to someone or something as being philosophical or not philosophical: philosophical meant reasoning and looking for the causes.

Herschel was a mathematician, chemist, and astronomer, the son of the famous astronomer William Herschel. Among other things, he discovered the action of hyposulfite of soda on otherwise insoluble silver salts in 1819, which led to the use of "hypo" as a fixing agent in photography. In 1839, he also invented a photographic process using sensitized paper. For Herschel, science was not merely collecting facts and specimens and organizing that knowledge into orderly "systems." True science was the search for causes and for general laws. This was the science of Newton and Bacon. One of his observations was that fields differed in their state of preparedness to be considered true sciences in the same league as mathematics and astronomy—that is, delivered of true causality and lawfulness. He thought that geology was prime to fall into line as the next true science.

The core of the book was Herschel's analysis of the nature of true scientific inquiry and his advocacy of Bacon's method of *vera causa* (true cause). In this approach, a cause is a true cause if it meets certain criteria. First, it must be a *real* phenomenon that could potentially act as a cause in the case under investigation. Second, this cause must be of the scale and nature that could in fact be *competent* to produce the observed result. Third, it must be demonstrated that this cause was in fact *responsible.* One can only too easily imagine cases in which a cause is ascribed to a phenomenon but is really an artifact. For example, it used to be thought that malaria was caused by miasmas of the air. The air in a swamp may be foul to breathe, and sickness demonstrably occurs in people who travel in such swamps. But the real cause is a microbe carried by mosquitoes.

The difficult part of vera causa is to prove *responsibility,* and in the case of something historical—that has already happened and will not be exactly repeated—it may be almost impossible. Experiments may show the role of the mosquito and microbe in malaria, but it will never be possible to be sure that a person who died in 1850 with a high fever actually had malaria and not septicemia. When Darwin came to write *On the Origin of Species,* he used the vera causa method to establish the case for natural selection, and hit the same problem.

ELEVEN

Geology Again

In addition to his reading and the controversies swirling about in the heady atmosphere of the world of science, Darwin had no shortage of more down-to-earth subjects to occupy him. In May of that year, his friend J. M. Herbert presented him (anonymously) with a Coddington's microscope. "It will give peculiar gratification to one who has long doubted whether Mr. Darwin's talents or his sincerity be the more worthy of admiration, and who hopes that the instrument may in some measure facilitate those researches which he has hitherto so fondly and so successfully prosecuted."[1] Coddington's newly invented microscope had a single lens but delivered about 360 power—a far cry from the sort of crude instruments that Grant had had in his laboratory at Edinburgh. And it was expensive, costing in the range of 5 or 6 guineas (equivalent to £400 or £500 today, or $800 to $1,000). This magnificent gift, a remarkable testament to friendship, launched Darwin back into a world that he had abandoned, under such unpleasant circumstances, four years before.

A second preoccupation was travel—or, rather, potential travel. After reading heavily in narratives of travel at Edinburgh and Cambridge, he was ready to make his own expedition to somewhere warm and exotic. "I have long had a wish of seeing Tropical scenery & vegetation."[2] He and a group of friends would follow in Humboldt's wake to a tropical place and study natural history; they would experience a version of nature that was totally different from the austere biological diversity and ancient vegetation-covered rocks of Europe.

Darwin worked on his dream with Henslow, who was, at least at first, an active enthusiast for the idea. The closest tropical paradise was Tenerife, with its magnificent mountain peak, Pico de Teide. And what could be better than to follow Humboldt's footsteps there, to the legendary "Dragon tree"? This ancient tree had been noticed by travelers since the fifteenth century, and Humboldt had been "not the less struck with its enormous magnitude. . . . Its height appeared to us to be about 50 or 60 feet; its circumference near the roots is 45 feet. . . . Among organic creations, this tree is undoubtedly, together with the Adansonia or baobab of Senegal, one of the oldest inhabitants of our globe." The dragon tree also presented "a curious phenomenon with respect to the migration of plants. It has never been found in a wild state on the continent of Africa. The East Indies is its real country. How has the tree been transplanted to Tenerife, where it is by no means common?"[3]

In addition to Darwin and Henslow, the members of the expedition were to be three other keen naturalists. Somewhat oddly, however, they were not contemporaries of Darwin. They were not students with an itch for one great adventure before they settled down but were Henslow's friends and, like Henslow, a dozen or more years older than Darwin. Marmaduke Ramsay was a tutor at Jesus College, William Kirby was a fellow of St. Clare's, and Richard Dawes was a tutor at Downing College. These were men who attended Henslow's soirees and were well established in the university's scientific elite, but Darwin hardly knew them personally. The choice of Ramsay, who was also a friend of Sedgwick, is particularly interesting because, as the ever-gossipy Darwin wrote to Fox, he was someone "whom I used formerly to dislike."[4] (In the same letter, Darwin said that he had finally come around to liking Leonard Jenyns more and more, but Jenyns was not invited.)

While Darwin continued to make plans for Tenerife, something cropped up to divert him. Henslow had been actively encouraging Darwin to take up seriously the subject that Herschel and others considered to be currently the most philosophical of the subjects in the natural sciences—geology. Geology was a very suitable subject for Darwin. In depending on correctly identifying and classifying rock types and their mineral composition, it had all the attractions of entomology in terms of collecting and accurately identifying and classifying data and specimens, it was intensely practical and suited to someone who loved to walk and climb, and it was not as oversubscribed among English clerics as were entomology and botany. It could be practiced anywhere; and it was a subject of rich intellectual promise. It was a young, modern, and increasingly popular science, entirely appropriate for a young man. The only possible disadvantage, from Darwin's point of view, was that it had become so controversial.

Here we come up against another of the curious inconsistencies in the *Autobiography*. Darwin said there: "I was so sickened with lectures at Edinburgh that I did not even attend Sedgwick's eloquent and interesting lectures. Had I done so I should probably have become a geologist earlier than I did."[5] This statement is, as the expression goes, either false or insufficiently true. Two fellow students recall Darwin's having attended Sedgwick's course.[6] (Perhaps, having earlier dismissed "lectures" as being utterly inferior to "reading" as a mode of learning, he could hardly bring himself to admit to having attended Sedgwick's lectures.) He did acknowledge that he had taken Henslow's botany course, but he was careful both to tell the reader that he "did not study botany" and to extol the pleasures of the field trips.

The only possible explanations for this inconsistency are either that Darwin had genuinely forgotten that he had attended Sedgwick's course—which seems unlikely given his later close association, on various levels, with Sedgwick—or that he was expressing the literal truth—he did not "take" the course formally, but he did attend at least some of Sedgwick's lectures. Thereby, just as he had discounted the value of Jameson's course at Edinburgh, he could avoid having to declare that he had learned anything of either botany or geology at Cambridge. He could leave the reader with the conclusion that five years at two major universities had taught him little but classics and logic.

Whatever the case, Darwin would need geology for his expedition to Tenerife. In particular, he needed to add some experience in practical field geology to the mineralogy and principles that he had learned from Jameson. Henslow took his education in hand and advised him to buy a clinometer (for measuring angles of strata and elevations), modified according to Henslow's specifications. Darwin went home to practice in Shropshire, first on tables arranged willy-nilly in his bedroom, then on the estate. He then decided to work on a geological map of the county—which was no small ambition for someone who supposedly knew little or no geology.[7] Soon he was writing ruefully to Henslow, admitting his failures: "I suspect the first expedition I take, clinometer & hammer in hand, will send me back very little wiser and a good deal more puzzled than when I started." It is an interesting letter because it shows that already he was thinking about geology in broad terms rather than minutiae. It continued: "As yet I have only indulged in hypotheses; but they are such powerful ones, that I suppose, if they were put into action but for one day, the world would come to an end."[8] Darwin was thinking of himself (and dared to express himself) in terms of being a natural philosopher rather than just an amateur collector.

Remarkably (at least if Darwin was as little schooled in geology as he claimed), in July Adam Sedgwick invited Darwin to go as his field assistant for a short study of the geology of north Wales. Evidently Henslow had pressed Sedgwick to take Darwin. It would be interesting to know just what Sedgwick thought Darwin could add to his work. Would he be a helper or simply a student? It is possible both to overestimate and to underestimate Darwin's role as an assistant to Sedgwick during the Welsh trip, but it seems improbable that Sedgwick would have taken with him a student who had not even sat in on his lectures, and who knew no mineralogy and could not identify rock types.

Sedgwick took Darwin to the Vale of Clwyd, where he wanted to investigate the reported presence of an extended layer of Old Red Sandstone (Devonian) in contact with the Carboniferous Limestone. The Old Red had been recorded on current geological maps (by G. B. Greenough) on the basis of a red sandy deposit. Sedgwick and Darwin took parallel routes on either side of the valley, then compared notes. And Darwin did not hesitate to venture the opinion that there was no Old Red. Together, they worked out a geological history of the valley as a simple syncline (down-warping of the earth's surface). Sedgwick, acting as tutor, next set Darwin to looking at the interesting question of the relationship between the planes along which slate cleaves to the original plane of deposition. Darwin had discovered that he had an ability as a practical geologist.

Darwin then struck off on his own, still making geological observations.[9] One day, he fell in with a young man who was also hiking in the mountains. Robert Land, the future Lord Sherbrooke, wrote of coming across Darwin: "He was making a geological tour in Wales, and carried with him, in addition to his other burdens, a hammer of 14 lbs weight. I remember he was full of modesty, and was always lamenting his bad memory for languages. . . . I saw something in him which marked him out as superior to anyone I had ever met: the proof which I have of this was somewhat canine in nature, I followed him. I walked twenty-two miles with him."[10]

The next month, Sedgwick wrote Darwin a long letter from Tremadoc in Wales, full of geological details and debating with Darwin some of the latter's observations concerning the slates of Cywm Idwal and Moel Shabod. "I don't agree with you in thinking that the mass of trap on ye crest of the hill is *under* the slate. It seems to me to be decidedly *over* it."[11] He had evidently come to see Darwin more as a junior colleague than a student.

The Welsh trip can be seen as yet another turning point for Darwin.[12] The young man whose inquiring mind had been churning away since childhood had acquired a great deal of knowledge and experience in the natural world. He had learned well from Jameson, Hope, and Grant in Edinburgh and Henslow and Sedgwick in Cambridge. But Darwin had no idea what would happen next. Would it be more adventures, or would he settle down in some quiet parish? Nonetheless, as far as the story of Darwin's life is concerned, at least a period of preparation did end, complete or not. Ready or not, he plunged into the next phase. And it was truly a plunge, taken like someone diving into an ice-cold swimming pool not sure where the bottom is.

TWELVE

HMS *Beagle*

Darwin's planned expedition to Tenerife and Humboldt's fabulous dragon tree was probably pie-in-the-sky—the sort of thing that sounds wonderful in theory but rarely works out in practice. In the real world, people have jobs, obligations, and a shortage of cash. Undergraduate dreams have a way of turning into a colder, harder reality after graduation. The expedition would not take place until June of the following year (1832) and, as the spring went by, the enthusiasm of some participants had already begun to flag. Even Henslow began to have serious doubts. Darwin could afford it, having a private income and no career. But even if Darwin could take the lion's share of the expenses, who would take care of Henslow's duties at home, to the university and not least his obligations to Mrs. Henslow and the new baby?

Darwin wrote to Henslow on July 11, "I am very anxious to hear how Mrs Henslow is.—I am afraid that she will wish me to the bottom of the Bay of Biscay, for having been the first to think of the Canaries."[1] A few days later, he wrote to Fox that he thought "Henslows chance of coming is *very* remote."[2] The final blow came when Darwin returned home from Wales in August 1831 to learn that Marmaduke Ramsay had died (from septicemia, as was so often the case). With that, it seemed that the much-wished-for expedition had surely slipped away.

Darwin must have been dumbstruck, however, to read a letter from Henslow that awaited him in Shrewsbury. After commiserating with Darwin on Ramsay's death, he plunged into the news of a fabulous opportunity, enclosing a letter from George Peacock (mathematician and astronomer at Trinity College, Cambridge) that explained matters more fully.

My dear Henslow, Captain Fitz Roy is going out to survey the southern coast of Terra del Fuego, & afterwards to visit many of the South Sea Islands & to return by the Indian Archipelago: the vessel is fitted out expressly for scientific purposes, combined with the survey; it will furnish therefore a rare opportunity for a naturalist & it would be a great misfortune that it should be lost.

An offer has been made to me to recommend a proper person to go out as a naturalist with this expedition; he will be treated with every consideration; the Captain is a young man of very pleasing manners (a nephew of the Duke of Grafton), of great zeal in his profession & who is very highly spoken of; if Leonard Jenyns could go, what treasures he might bring home, as the ship would be placed at his disposal, whenever his enquiries made it necessary or desirable; in the absence of so accomplished a naturalist, is there any person whom you could strongly recommend: he must be such a person as would do credit to our recommendation.[3]

Henslow told Darwin: "I consider you to be the best qualified person I know who is likely to undertake such a situation—I state this not on the supposition of yr. being a *finished* naturalist, but as amply qualified for collecting, observing, & noting any thing worthy to be noted in Natural History. . . . Capt. F. wants a man (I understand) more as a companion than a mere collector & would not take any one however good a Naturalist who was not recommended to him as a *gentleman*. . . . you will have ample opportunities at command—In short I suppose there never was a finer chance for a man of zeal & spirit."[4]

We are so used to the fact that Darwin went off on the *Beagle* adventure that we sometimes forget to wonder about the sheer implausibility of it. It was quite extraordinary that Darwin, who to this point had always needed the support of friends and relatives, was ready to fledge the nest so completely. He would be forced to rely on himself, something he had signally failed to do so far. There would be no father, no sister, no brother Erasmus, no Henslow—just this unknown Captain FitzRoy, who was evidently an aristocrat rather threateningly connected to the very social order that Henslow and his circle were currently fighting over the Reform Bill.

Darwin, however, instantly saw that it was an opportunity—however fraught with difficulty and even danger—that could not be passed by. It may be true that for Darwin this was an offer from heaven as, at that particular moment, he had not the slightest idea what he would do next. He knew only that he had a deep commitment to a scientific life. With this offer, his undergraduate dream of an expedition to the tropics had metamorphosed into a major adventure. But it was crazy. His father would disapprove the whole idea.

How much did Darwin worry about the practical exigencies of the expedition—two years (as was then planned) on a navy ship manned presumably by the usual sorts of ruffians, and heading to parts heathen and dangerous? It would mean bad food, loneliness, possibly danger, certainly seasickness . . . no one in his right mind would accept such an offer. Except, as Peacock wrote and Henslow echoed, it was an opportunity not to be lost. Darwin later discovered that Jenyns, despite having the responsibility for two livings, "was so near accepting it, that he packed his clothes. . . . Henslow himself was not very far from accepting it: for Mrs. Henslow, most generously & without being asked gave her consent, but she looked so miserable, that Henslow at once settled the point."[5]

Although circumnavigations were very rare, this voyage of HMS *Beagle* (her second) had been planned explicitly to make a complete circle of the globe. Its primary purpose was to make a circle of measurements of longitude using the same instruments and the same observers. It would be the first circumnavigation of its kind and immensely valuable in standardizing maps and navigational instructions. If successful (that is, if longitude could be measured accurately), the exact coordinates of major ports of call around the globe would be fixed. Such consistency was badly lacking. Beyond that, FitzRoy was planning a fully scientific cruise. Even if the plans had been more mundane, the opportunities for collecting new species at a succession of new and strange places would be unparalleled.

A second letter arrived from Peacock confirming his first and stating (perhaps overoptimistically): "The expedition is entirely for scientific purposes & and the ship will generally wait your leisure for researches in natural history &c: Captain FitzRoy is a public spirited & zealous officer, of delightful manners & greatly beloved by all his brother officers: . . . [he] spent 1500£ in bringing over & educating at his own charge 3 natives of Patagonia: he engages at his own expenses an artist at 200 a year to go with him: you may be sure therefore of having a very pleasant companion, who will enter heartily into all your views."[6]

Few fathers would have approved (and financially supported) such a scheme. Dr. Darwin naturally said no. His objections to the proposal were perfectly sound: it was inappropriate for someone intending to settle down as a clergyman; Darwin had no experience of the sea; the notice was too short; there was every chance that Darwin would not find FitzRoy compatible. Furthermore, "what was the reason, that a Naturalist was not long ago fixed upon?"[7]

Falling from the emotional heights to the depths, Darwin agreed to reject the offer. Then he set about finding ways to turn his father around, which he did through the loophole that the good doctor (either inadvertently or well-intentionedly) had left open. Robert Darwin was always willing to be swayed by his brother-in-law, whom he knew to be exceptionally fond of Darwin and a man of great common sense. Dr. Darwin wrote to Josiah Wedgwood about the offer: "I strongly object to it . . . if you think differently from me I shall wish him to follow your advice."[8] Presumably he was confident that Josiah would also say no. But Darwin rode off to Maer and readily persuaded his Uncle Jos.

Josiah Wedgwood replied to Robert Darwin immediately. He did not think that the voyage would be something disreputable to a clergyman, as natural history, "though certainly not professional is very suitable." The

Admiralty wouldn't send out a ship that wasn't safe. The experience would be good for Darwin, perhaps help him to settle down. The voyage would be of no use to him in terms of a profession, "but looking upon him as a man of enlarged curiosity, it affords him such an opportunity for seeing men and things as happens to few."[9]

The next step was more tricky. Darwin had to be approved by Captain FitzRoy and, in turn, Darwin had to be persuaded that the venture really would be everything it was advertised to be.

Although it was not uncommon for a naval ship to have on board several supernumeraries, FitzRoy planned to take an unusual number, especially for a vessel as small as the *Beagle*: an instrument maker to tend the chronometers, an artist, a fiddler for the poop cabin, a missionary, and even three natives of Tierra del Fuego. But FitzRoy, as we will see, was of a very scientific disposition, and he also had to have a naturalist. A long expedition of exploration like this would greatly profit from having a full-time naturalist on board. He would have a lot in common with the captain, who was himself fascinated by everything in science.

The captain was offering to share his cabin for all meals, although the naturalist would sleep elsewhere. For this arrangement to be remotely possible, two conditions had to be met. The man would have to be someone not in the service (the responsibilities of command required the captain to hold himself socially apart from the other officers and men). In order to share FitzRoy's table and FitzRoy's confidence, this person would also have to be a thoroughgoing gentleman. FitzRoy, having spent all his adult life at sea, did not know many naturalists and indeed had a rather small range of civilian acquaintances, very few if any of whom would be free to drop everything for a round-the-world voyage. So when FitzRoy placed his request to take a naturalist on board with his superior, Captain Francis Beaufort (head of the Hydrographic Survey Office), he had also asked Beaufort to find him the right man.

The answer to Robert Darwin's question of why a naturalist had not been chosen long before is that FitzRoy's orders had only been issued on June 27. FitzRoy had returned to Britain as captain of HMS *Beagle* on its first voyage the previous year. The *Beagle* had been sent to take part in a survey of the coasts of South America as tender to a large brig, HMS *Adventure* (commanded by Philip Parker King).[10] The work was supposed to concentrate on the Straits of Magellan and there, under terrible conditions, the *Beagle*'s captain, Lieutenant Pringle Stokes, had committed suicide. FitzRoy, then in the favored position of flag officer to the admiral of the South American Station, had been put in charge and the *Beagle* sent back south. Over the next year, further difficulties plagued the ship, particularly harassment by the Fuegian Indians, who stole whatever they could, including boats. FitzRoy took four Fuegians hostage and, greatly to the Admiralty's surprise and displeasure, brought them back to England.[11] The plan seems to have been to educate them and return them to Tierra del Fuego, where they would convert their fellows and become an asset to visiting sailors.

FitzRoy had immediately begun canvassing the Admiralty for a second commission but at first it refused. In desperation, FitzRoy hired a brig to sail back south. Then, at the last minute, the Admiralty (probably under pressure from FitzRoy's uncle Lord Londonderry) commissioned the new and extremely ambitious voyage.

Much has been made of FitzRoy's motives in taking a naturalist on this voyage. Did FitzRoy really want a naturalist or was he simply looking for a gentleman companion to ease the hardships of a long, lonely voyage? Did Darwin really qualify as a naturalist? Some authors have seized upon Henslow's phrase "more as a companion" to argue that Darwin was taken on simply as the captain's amanuensis.[12]

One important question is: what did *Darwin* think his role would be? Clearly, Darwin thought he was to be taken on wholly as a naturalist. Everything he did confirms that point. He was at great pains to negotiate with Beaufort so that he kept control over his collections. Throughout the negotiations (and indeed through the voyage), Darwin insisted that he would go only if the cruise included a full circumnavigation. He must see the South Seas; South America alone (where the previous *Beagle* expedition got bogged down) would not be enough.

All this is a confirmation of Darwin's status as a serious naturalist. However, the scientifically inclined person FitzRoy wanted for the voyage would also distinctly be his companion, rather than an employee, and would occupy a special place on board. There may also be some truth in the assertion that FitzRoy was already worrying about the extreme responsibility of a solo voyage around the world, his own sometimes unstable personality, the fact that Pringle Stokes had killed himself during the ship's first voyage to Cape Horn, and the prevalence of suicide in his own family.

A close reading of the letters clarifies the issue.[13] A consistent theme in Peacock's and Henslow's letters to Darwin is that FitzRoy was a gentleman, not some rough seaman, and that FitzRoy's offer to take a naturalist required that the naturalist be equally a gentleman. At that time, the fashion for natural history collections had created a category of naturalist collectors who fanned across the globe collecting specimens for wealthy patrons. None of these were men of any social standing. They were the "mere collectors" that Henslow specifically stated FitzRoy did not want. What was needed was someone "recommended to him as a *gentleman*" (Henslow's emphasis). Similarly, the Cambridge dons would not send someone off with FitzRoy and all those uncouth naval types unless they were assured that he would be treated as a gentleman. Darwin could not possibly have accepted the appointment on any other terms.

In other words, Peacock and Henslow had emphasized the term "companion" because they were assuring Darwin that it would be socially and personally acceptable for him to go. FitzRoy really did want a naturalist, but someone with whom he could be on equal terms during the long voyage. This is confirmed by the fact that Jenyns and Henslow considered the appointment for themselves. Eventually, as a result of his inquiries at Cambridge, Beaufort wrote back to FitzRoy: "Mr Peacock . . . has succeeded in getting a 'Savant' for you—a Mr Darwin grandson of the well known philosopher and poet—full of zeal and having contemplated a voyage on his own account to S. America."[14]

With the word "savant," Beaufort summed up the situation. In the last two-thirds of a year at Cambridge, Darwin had found himself intellectually. The key to that intellectual strength was not the amassing of knowledge per se but confidence with, and about, logical systems. Humboldt, Paley, and Herschel, each in quite different ways, had given him intellectual confidence. He had always had the physical ability and the social skills if he wanted to apply them. He could be tough and charming, equally. He could ride, shoot, and put up with hardships. What he needed was a reason to do so.

Darwin was coming to see himself as someone with a career to make, a reputation to establish. He wasn't at all sure what his particular contribution to science would be—there is no hint of him having been a closet transmutationist at that time. But he knew he would be something. In other words, he felt a "calling" exactly like the one that had been missing in his preparations for the church.

But, after some six years of formal education at Edinburgh and Cambridge, and an even longer informal period as a keen hobbyist, what kind of "savant" was Darwin? Darwin cannot be dismissed as some amiable but aimless amateur, who had only been whiling away the days before retiring to some sinecure deep in the English countryside. Certainly, Henslow and Peacock would not consider someone who had just graduated as a "poll man" to be a scholar. Henslow's phrase "not a *finished* naturalist" meant that he lacked the sort of breadth and depth of experience that Peacock and Jenyns (or Henslow himself) had. He had not traveled outside Britain. Apart from his deep experience in entomology, most of his knowledge of natural history was from reading and lectures. That experience was extensive, however.

He had learned geology and mineralogy from both Jameson and Sedgwick, botany from Henslow. In entomology and ornithology, his knowledge was far beyond most, if not all, his contemporaries. Even more important was Darwin's habit of mind. He may not yet have been a scholar, but he had strong intellectual aspirations. Beyond the facts, he was interested in principles, processes, and causes. As he had written to Henslow in July, his geological hypotheses were "such powerful ones" as to bring the world to an end.[15] The Charles Darwin who applied himself seriously to science in the spring of 1831 was a man who, as he had long imagined, would accomplish great things.

It was now up to FitzRoy and Darwin to find out if they could get along together. Robert FitzRoy was in many ways Darwin's opposite. A Tory from an aristocratic family, he was short, dark, self-possessed, and worldly. He was descended from the illegitimate son of King Charles II by his mistress Barbara Villiers. The king eventually settled upon the family the title Duke of Grafton.

Nothing could have been further from the Whig Darwins and Wedgwoods. Yet, in many ways, FitzRoy and Darwin were very similar. Both were the second sons of men who were also second sons and had suffered their whole life under the shadow of their great relatives. FitzRoy's grandfather was the Duke of Grafton, one uncle was Lord Castlereagh, another was Lord Londonderry. Both FitzRoy and Darwin had lost their mothers at an early age and were brought up by sisters and then sent off to boarding school (the Royal Naval College in FitzRoy's case—he graduated top of his class). Both were depressives (FitzRoy would now be termed manic-depressive or bipolar). Both had uncles who were suicides. Both tended to dwell excessively on matters of health. Above all, both were extremely intelligent and well read. FitzRoy, like Darwin, was a keen scientist, but much more mathematically gifted. In 1831, FitzRoy was twenty-six and Darwin twenty-two years old.

On both sides, it was an odd situation. Darwin and FitzRoy would be spending a lot of time together, yet not only were they strangers, neither had chosen the other. As a naval officer, FitzRoy expected to be thrown together with strangers and knew how to make a good job of it. For Darwin it was a tougher decision.

FitzRoy was a scientist manqué. It was immediately after his first long conversation with FitzRoy that Darwin wrote to his sister Susan back at home for various scientific papers: Humboldt on isothermal lines, Coldstream and Foggo on meteorological observations, and so on (see chapter 7). The timing suggests that Darwin and FitzRoy had been discussing the latter's long-standing interest in the use of the barometer for forecasting changes in weather (an interest he shared with Robert Jameson).

Now time was of the essence. The expedition was due to leave at the end of October. FitzRoy's approval of Darwin was essential in every sense. But so was Darwin's approval of FitzRoy. And now it is clear that the old diffident and cautious Darwin of Edinburgh and Cambridge days was being replaced by someone more confident, if naive. After that first meeting, Darwin was euphoric: "He is my beau ideal of a captain," but he took the precaution of asking an acquaintance, Alexander Wood, who was a great friend of FitzRoy, for a view of the situation. Wood replied positively to Darwin, but then Wood got news from FitzRoy "so much against my going, that I immediately gave up the scheme." FitzRoy had developed cold feet about taking on board "someone he should not like."[16]

When they met again, FitzRoy made the excuse that an offer had been made to someone else. As he was no longer able to go, the offer to Darwin was back on the table. Darwin, almost desperate that everything should be satisfactory, wrote both to his sister Susan and to Henslow, exclaiming at how wonderful he thought FitzRoy was. "You cannot imagine anything more pleasant, kind & open than Cap. FitzRoy's manners were to me.—I am sure it will be my fault, if we do not suit." FitzRoy, in turn, was at pains to tell Darwin that the ship would be dreadfully cramped, "I must live poorly, no wine & the plainest dinners." But if at any time Darwin wanted to leave, then he could. Darwin reported that "[the] stormy sea is exaggerated. the manner of proceeding will just suit me. They anchor the ship & remain for a fortnight at a place."[17] Perhaps the clincher for Darwin was the fact that FitzRoy was planning to take a huge library, and his enthusiasm for taking along enough scientific books was exceeded only by his insistence on taking enough firepower, in terms of shotguns, rifles, and pistols. If FitzRoy had been conducting a psychological campaign to win Darwin over, he could scarcely have chosen better arguments.

On September 10, Darwin and FitzRoy made their first voyage together, on a packet steamer—no seasickness!—from London to

Plymouth, where they found the *Beagle* in Devonport Dockyard. It hardly had the appearance of a gleaming, refurbished survey ship, however: "without her masts or bulkheads & looked more like a wreck, than a vessel commissioned to go around the world."[18] They would not be going anywhere for quite some time.

Darwin returned to London to continue his preparations. He had a dizzying amount of work to do and purchases to make (on the doctor's account): books, notebooks, specimen boxes, preservatives, guns, ammunition, collecting nets, labels, pencils, waterproof gear, india ink, pen nibs, knives, a telescope—all the paraphernalia he would need for a life in exile. Once they left England, anything left behind might take months to arrive, if it did at all. Sisters and cousins all pitched in to help Darwin prepare, as did the ever-faithful Henslow. It was a tremendously exciting time.

On October 24, Darwin took a coach back to Plymouth, where he found the *Beagle* looking a bit more like a vessel that might actually weather Cape Horn and again waxed enthusiastic. "Our cabins are fitted most luxuriously with nothing except Mahogany: in short, every thing is going on as well as possible. I only wish they were a little faster."[19]

There were already some tricky moments with FitzRoy; the man turned out to be a martinet with his crew and was very unpredictable as a companion—one minute a jolly friend, the next a sour, distant critic and an appalling snob. He could fall into dreadful, irrational rages; he could be devastatingly charming. But currently FitzRoy was preoccupied with the ship, leaving Darwin to worry about the lack of space. Where he would fit everything? Where would he fit himself?

At over six feet in height, he was not built for ships, especially this one. He had a small cupboard in the forecastle for his scientific gear but nothing like a cabin for himself. His sleeping space was merely a hammock slung in the main poop cabin, over the great table where the officers would be drawing up their charts. Of privacy there was none. He would even have to share the cabin with two midshipmen. In order to be able to sling a hammock for Darwin, it was necessary each night to remove one of the bookshelves.

All this would have been off-putting to the most hardened traveler, let alone the rather cosseted Darwin. We have little idea of *exactly* how he felt. But, as the days of delay dragged into months, he sank into a deep depression. In this period, his letters home stopped and his health changed. The worse he felt, the more he worried about it, compounding the problem. His hands were inflamed and the most worrying symptom was new: heart palpitations. He was sure that he was going to have a heart attack, or was actually having one. But he said nothing for fear that his father would summon him home and the opportunity of the voyage would be lost.

As we look at these first few months, first stuck in Plymouth and then finally at sea, something of the character of the two men comes through. First, it is clear that FitzRoy very quickly became the new version of a mentor for Darwin, but not so much in the sense that Grant or Henslow had been, or even Erasmus. Darwin deeply admired FitzRoy and yet was already a little nervous of his erratic disposition. And with good reason. But FitzRoy was an enthusiast, interested in everything, and he shared as much as time allowed in Darwin's work. He was also an enormously authoritative figure and, like most authorities, someone to be agreed with in public and questioned only in private, and then very carefully. Darwin later wrote of FitzRoy, "He is a very extraordinary person. . . . His greatest fault as a companion is his austere silence: produced from excessive thinking: his many good qualities are great & numerous: altogether he is the strongest marked character I ever fell in with."[20] And in turn FitzRoy confided to Beaufort, "Darwin is a very sensible, hardworking man, and a very pleasant messmate. I never saw a 'shore-going' fellow come into the ways of a ship so soon and so thoroughly as Darwin. I cannot give stronger proof of his good sense and disposition than by saying 'Everyone respects and likes him.' "[21]

Darwin very quickly became on good terms with the other officers, who were a very intelligent, keen bunch, not at all the ruffians that he might have expected (the seamen were another matter). With these men, all much the same age as himself or younger, he was an important person: an expert. They quickly came to call him "Philos." He may have known nothing about ships and the sea, but he established himself at once as a man of scientific authority. He was likeable and popular, joking that the inside of a ship was like the inside of a stomach, "a large cavity containing air, water & food, mingled in hopeless confusion."[22] When he needed shelter from FitzRoy's tantrums, the other officers took him in.

Darwin's enthusiasm for natural history soon became infectious. They all became collectors. Wickham, the first lieutenant, who had the difficult job of dealing with the messes that Darwin's collecting made on the holy and holystoned decks of the *Beagle,* had his own name for Darwin. He called him, affectionately, "that damned flycatcher." The mate, Bartholomew James Sulivan (later Sir Bartholomew), became a friend for life.

Only one man, Robert McCormick, struck a jarring note. And the reasons were very simple. McCormick, as ship's surgeon, by office held the formal post of ship's naturalist. It was one of the many roles of the surgeon on a king's ship, but not a major one on most ships. He was expected (when and as time permitted) to collect specimens and to have a sharp eye out for items of economic importance like mineral ores. Many naval surgeons had wonderful scientific careers—including two men who would later be very important in Darwin's life: Joseph Hooker (HMS *Erebus*) and Thomas Henry Huxley (HMS *Rattlesnake*).

McCormick had more serious aspirations to make a name for himself in natural history than did most ship's surgeons. In a terrible irony, to prepare himself while waiting for the right ship, McCormick had spent a year at Edinburgh studying with the best man available—none other than Professor Robert Jameson! He had taken Jameson's course expressly for the purpose of preparing himself should an opportunity like this arise. How desperately McCormick must have cringed at seeing Darwin arrive on board, feted as a "philosopher" and enjoying the captain's favor—as *the* naturalist.

Given that both men had prickly personalities and were extremely ambitious, strife between Darwin and McCormick was almost inevitable. Both had too much at stake to have any intention of sharing the experience with another. Before the ship had left port Darwin was already sniping at McCormick. In letters home, he wrote: "My friend the Doctor is an ass, but we jog on very amicably: at present he is in great tribulation, whether his cabin shall be painted French Gray or a dead white—I hear little excepting this subject from him."[23] More was to follow.

Benjamin Bynoe, the assistant surgeon of the *Beagle*, turned out to be no mean naturalist himself. A warm and generous man, he never felt threatened by Darwin. In fact, he and Darwin worked closely together—which is to say that he assisted Darwin. (During the succeeding voyage of the *Beagle*, Bynoe would make his own important collections of Australian birds and conduct important investigations into the reproduction of kangaroos, based on field dissections.)[24] Having seen the relationship between FitzRoy and Darwin, Bynoe realistically made cooperation, rather than opposition, his means to a happy and useful voyage. He probably learned a great deal from associating with the immensely better-read and well-informed Darwin. But their friendship was based on more than that, and without Bynoe's careful nursing, Darwin might never have recovered from a serious bout of illness in Chile in 1835.

THIRTEEN

Epiphanies

Many themes can be traced out in the nearly five years of the voyage of the *Beagle*. Darwin's relationship with Robert FitzRoy provides the greatest personal story, although this has become thoroughly confused because of an article by Darwin's granddaughter Nora Barlow (otherwise one of the most accurate and original of Darwinian students) published in 1932, that still has a popular currency. In this essay, she suggested that one of the causal factors in Darwin's intellectual development in general and the formulation of his theory of evolution in particular was the opposing natures of Darwin and FitzRoy.[1] Specifically, she saw an ongoing debate between Darwin the radical atheist and FitzRoy the reactionary Christian. It is an appealing notion: a clash of cultures, a dynamic tension. On the contrary, however, during the voyage, and particularly at the beginning, Darwin was the more conventionally religious of the two (he claims to have been laughed at by the *Beagle*'s officers because of his belief in the literal truth of the Bible). FitzRoy, by his own admission, was something of a freethinking agnostic, carried away by the ideas of people like Rousseau.

It is true that FitzRoy later became a religious fundamentalist, but that change happened only after the voyage had ended, and probably under the influence of his extremely religious wife, Mary. FitzRoy "got religion" late and quickly. Darwin lost his faith later and very slowly. What the two men shared was a fascination for science and exploration, entering fully into each other's enthusiasms and forming a close, if erratic, bond of friendship. During the voyage, Darwin matured greatly and, as his confidence in his own abilities and potential grew, so did his independence. His growing distance from FitzRoy (once his beau ideal) was not just a matter of FitzRoy's difficult personality (personalities), but was also due to the fact that Darwin no longer needed to lean on someone else.

Another constant theme of the voyage was Darwin's seasickness, which began before they had even left Plymouth and, perhaps uniquely in maritime annals, continued until the day he reached English soil once more.

The start of the voyage could not have been less propitious. In a season of steady southeast gales, it was impossible for the tiny *Beagle* to leave port. On two occasions, having tried, FitzRoy had to put back into port. On one occasion, the ship grounded on Drake's Rock, right in Plymouth Sound.[2] The weather improved, and they could have sailed on Christmas Day, were it not for the fact that the crew was mostly incapable through drink.

When the *Beagle* finally sailed south from Plymouth on December 27 in a perfect easterly breeze, Darwin was utterly seasick. All his previous optimism from having survived the steamer to Plymouth in October evaporated; it went overboard, so to speak. With a voyage of at least two years ahead of him, Darwin's mental misery must have been as appalling as the nausea, and of course would have fed directly back into the seasickness he was trying to control. The sight and sound of the crew being flogged for their Christmas sins did nothing to ease Darwin's mind. All he could do was lie in his hammock and read. As his mal de mer never really went away, even after a voyage that stretched into five years, Darwin got a great deal of reading done.

When the ship approached Tenerife, Darwin's spirits lifted. At last, the goal of the expedition that he had planned with Henslow was finally within reach. He would see Humboldt's famous tree. Alas, quarantine regulations prevented them from landing.

Confined to his hammock once more, Darwin started to read his copy of Lyell's *Principles of Geology* (the first volume).[3] Before the *Beagle* sailed, Henslow had advised Darwin to read Charles Lyell's much-talked-about book but "on no account to accept the views therein advocated."[4] On the *Beagle* there was no one to create a controversy around Lyell's ideas, certainly not FitzRoy, who had given him the book in the first place. Darwin was free to form his own opinion of the book on its own merits, and to do so without stress. It proved to be a revelation, giving him a new confidence in his intellect and a sense of direction in his life.

Lyell's erudite introductory survey of the history of geological ideas put Darwin's previous learning, and the Wernerian view of the world, into context. Werner and his students and disciples had done a superb job of setting out the basic arrangement of the geological column, following Daubenton and Buffon's division of rocks into the Primary and Secondary. But when it came to explanations, histories, and theories of the earth, the narratives that Wernerians preferred were preconditioned by their religion.

Lyell had been vilified at Cambridge because he removed religious premises from geological discussion and for his failure to discover evidence of the linearity in earth history implicit in the Judeo-Christian tradition. As he read on, Darwin saw how Lyell, in his logical way, cut through these older approaches and let the evidence speak for itself. In fact, Lyell followed the principles of Bacon's (and Newton's) vera causa, which Darwin had admired so much when set out in Herschel's *Preliminary Discourse*. His systematic approach to the data, unencumbered by any religious imprint, allowed the evidence of the rocks to direct the explanations.

Lyell's view of earth history was neatly captured by the illustration forming the frontispiece to the first volume. It was a drawing of Roman columns from the market ("Temple of Serapis") at Puzzuoli, near Naples. These white columns of limestone showed darker bands near their base that on inspection turned out to have been made by marine boring clams. Remnants of their shells were still in place. The only explanation was that these columns, and the dry land on which they had been erected, had been depressed below sea level, raised up, then depressed, then raised again. Given that the limestone itself had been formed on an ancient sea floor and then raised into the mountains from which the columns had been quarried, it was a marvelous demonstration of Lyell's concept of the dynamics of earth history.

What happened next was pure serendipity. At St. Jago in the Cape Verde Islands, their next landfall, Darwin went ashore with Lyell's revised view of geology in his head. When he examined the sea cliffs of Quail Island, Darwin identified a succession of layers: the remnants of an old volcano, subsidence, deposition of a calcareous layer with modern shells, then newer lava, and then subsequent elevation. Darwin had found that not only could he identify rocks (as he had with Sedgwick in Wales), he had the skill to read their history just as Lyell had shown him. After three days, he came back on board with a new vision of himself; he had discovered a vocation—as a geologist. He wrote a cheerful letter to his father: "I think, if I can so soon judge.—I shall be able to do some original work in Natural History."[5]

As the voyage continued, everywhere he traveled in South America, Darwin saw evidence that he could interpret only as indicting a relatively recent elevation of the whole land. He became more and more convinced that not only could he make significant contributions to understanding the geology of South America, he would write a book on the subject. This was quite a development for someone who had supposedly been a total novice in the subject twenty-four months before.

We have a clear image of Darwin the collector—the fishing nets, butterfly nets, fossils, guns, thousands of pinned insect specimens and dried plants. We do not usually see Darwin as a geologist, although that is how he first made his mark as a scientist. Darwin did not suddenly devote the rest of the voyage to geology, however. Not only did he have a duty to spend a great deal of time collecting animal and plant specimens, he was passionately interested in seeing for himself, observing, and collecting the extraordinary biological diversity of the tropics.

Accordingly, his next epiphany came in the tropical rain forest, which they encountered at their first landfall.

The *Beagle* put into Bahia de Todos los Santos (now San Salvador), Brazil, at the end of February 1832, and it was here that Darwin had his first, overwhelming experience of the tropical rain forest. Everything about Bahia enchanted him: "The town is fairly embosomed in a luxuriant wood & situated on a steep bank overlooks the calm waters of the great bay of All Saints . . . but these beauties are as nothing compared to the Vegetation: I believe from what I have seen Humboldt's glorious descriptions are & will for ever by unparalleled: but even he with his dark blue skies & the rare union of poetry with science, which he so strongly displays . . . falls far short of the truth."[6]

In his diary, Darwin wrote of

transports of pleasure: I have been wandering by myself in a Brazilian forest: amongst the multitude it is hard to say what set of objects is most striking; the general luxuriance of the vegetation bears the victory, the elegance of the grasses, the novelty of the parasitical plants, the beauty of the flowers—the glossy green of the foliage, all tend to this end.—A most paradoxical mixture of sound and silence pervades the shady parts of the wood, the noise from the insects is so loud that in the evening it can be heard even in a vessel anchored several hundred yards from the shore. Yet within the recesses of the forest when in the midst of it a universal stillness appears to reign. To a person fond of natural history such a day as this brings with it pleasure more acute than he ever may again experience. After wandering about for some hours, I returned to the landing place. Before reaching it I was overtaken by a Tropical storm. I tried to find shelter under a tree so thick that it would never have been penetrated by common English rain, yet here in a couple of minutes, a little torrent flowed down the trunk. It is to this violence we must attribute the verdure in the bottom of the wood, if the showers were like those of a colder clime, the moisture would be absorbed or evaporated before reaching the ground.

The next day's entry continued: "I can only add raptures to the former raptures. I walked with the two Mids [midshipmen] a few miles into the interior. The country is composed of small hills & each new valley is more beautiful than the last.—I collected a great number of brilliantly coloured flowers, enough to make a florist go wild.—Brazilian scenery is nothing more nor less than a view in the Arabian Nights, with the advantage of reality.—The air is deliriously cool & soft; full of enjoyment one fervently desires to live in retirement in this new & grander world."[7]

Darwin never lost his ability to combine a sense of wonder with scientific detachment and keen observation as he experienced new landscapes across South America. Later in the voyage, he used some of his more evocative language to describe geological scenes. In the Andes, he wrote: "As we ascended the valley the vegetation becomes exceedingly scanty; there were, however, a few very pretty Alpine plants. Scarcely a bird or insect was to be seen.—The lofty mountains, their summits marked with a few patches of snow, stood well separated one from the other . . . the bright colours, chiefly red & purple, of the utterly bare & steep hills;—the great & continuous wall-like dykes; the manifest stratification, which, where nearly vertical, causes the wildest & most picturesque groups of peaks."[8] And: "Excepting the Vultur aura, which feeds on the Carcases, I saw neither bird, quadruped, reptiles or insect. On the coast mountains at about 2000 ft. elevation, the bare sand was in places strewed over with an unattached greenish Lichen, in form like those which grow on old stumps: this in a few spots was sufficiently abundant to tinge the sand when seen from a little distance, of a yellowish colour. I also saw another minute species of Lichen on the old bones. And where the first kind was lying, there were in the clefts of the rocks a few Cacti. These are supported by the dense clouds which generally rest on the land at this height. Excepting this, I saw no one plant. This was the first true desert I have ever seen."[9]

FOURTEEN

Storms and Floods

As the voyage unfolded, Darwin would encounter situation after situation that challenged or changed his worldview, both in terms of sciences and of human affairs. Every place they came to was different, revealing new aspects of geology and new surprises of biological diversity. But while one naturally looks for the big events that, during the voyage, shaped Darwin's thinking, it cannot be emphasized enough that it was also the daily observations and constant collecting of specimens and the firsthand appreciation of a huge range of landscapes and environments (natural and human) that gave him the personal experience and the empirically based viewpoints from which, once back in England, he was then preeminently qualified to theorize. His great ideas were always grounded in a mastery of the detailed facts.

Fortunately for Darwin, each landfall was separated from the next by a sea passage, during which he could arrange his specimens and collect his thoughts. He kept up his diary religiously and tried to be meticulous about labeling his specimens, although he developed a bad habit of giving each a number referring to a notebook entry, rather than a full label that would stay with the specimen itself.

The peoples and institutions that he met as the ship worked its way down the South American coast were no less fascinating than the natural history. In their first year in South America alone, they encountered revolutions; all too many dictators, warlords, slavers, and slaves; colonists; gauchos; colonial administrators; farmers and peasants; and indigenous peoples. As the voyage wore on, they became policemen at the Falklands and reluctant missionaries in Tierra del Fuego. Crossing the Indo-Pacific, they negotiated on behalf of the crown with the queen of Tahiti; encountered idyllic south sea islands, Maori cannibalism, settlers, and convicts; the *Beagle* and its crew even played a tiny role in the extermination of the Tasmanian native peoples.

When the expedition reached South America, whenever the ship was in harbor for any length of time, or when they were working slowly along a coast with the officers busy with their surveying work, Darwin would, with great relief, escape. He would go ashore, either to a rented cabin or to explore on horseback, far away from the restricted space and equally confining personal intersections of the ship. In developing this pattern of working, Darwin was (intentionally or not), following one of Humboldt's dictates for effective exploration: "During long sea-voyages, a traveler hardly ever sees land, and when the land is seen after a long wait it is often stripped of its most beautiful products. Sometimes, beyond a sterile coast, a ridge of High Mountain covered in forests is glimpsed, but its distance only frustrates the traveler. It is not by sailing along a coast that the direction, geology and climate of a chain of mountains can be discerned."[1]

After the ship's first landfall in South America, the expedition went on to Rio de Janeiro, where it remained for three months. Darwin moved on shore, renting a cottage on Botofago Bay along with FitzRoy's artist, Augustus Earle, whose health was suffering badly from the dampness of the ship. With them also was one of the Fuegian hostages, Fuegia Basket, perhaps because she would be in a more controllable environment. Returning to the ship one day after a jaunt into the interior, Darwin discovered, "During my absence several political changes have taken place in our little world. Mr McCormick has been invalided, & goes to England by the Tyne." Bynoe was promoted to surgeon. Darwin wrote to Henslow: "As for the Doctor he has gone back to England.—as he chose to make himself disagreeable to the Captain."[2] To his sister Caroline he wrote: "He is no loss."[3]

McCormick later wrote in his autobiography: "Having found myself in a false position on board a small and very uncomfortable vessel, and very much disappointed in my expectation of carrying out my natural history pursuits, every obstacle having been placed in the way of my getting on shore and making collections, I got permission from the admiral in command of the station here to be superseded and allowed a passage home."[4]

The situation for poor McCormick had been untenable from the start, and he did not help matters by adopting an attitude of aggrievement and entitlement. Once Darwin started collecting—even before their first landfall, FitzRoy allowed him to drag a net behind the ship— McCormick saw how the land lay. Darwin really would have the lion's share of any natural history collecting, and FitzRoy intended to help by arranging the ship's schedule to suit him.

On St. Jago, Darwin took several walks with McCormick but found him a poor companion and very old-fashioned in his views. In that way he had of dismissing lesser mortals, Darwin referred to McCormick as a "philosopher of rather an antient date . . . at St Jago by his own accounting he made general remarks during the first fortnight & and collected particular facts during the last."[5] Presumably, McCormick also had a religious viewpoint on geology quite different from Lyell's. At Plymouth he had attended the dockyard chapel faithfully, twice every Sunday.

Darwin seems to have behaved quite high-handedly toward the ship's surgeon. When they landed at St Paul's Rocks, FitzRoy and Darwin, in high spirits, set about collecting birds. Reports had said that the boobies and terns were so tame that one could kill them with a geological hammer, and FitzRoy and Darwin did just that. FitzRoy wrote, "The first impulse of our invaders of this bird-covered rock, was to lay about them like schoolboys; even the geological hammer at last became a missile. 'Lend me the hammer?' asked one. 'No, no,' replied the owner, 'you'll break the handle,' but hardly had he said so, when, overcome by the novelty of the scene, and the example of those around him, away went the hammer, with all the force of his own right-arm."[6] Darwin in his own diary makes it clear that it was he who had the geological hammer. "We knocked down with stones & my hammer, the active and swift tern. Shooting was out of the questions, so we got two of the boat's crew & the slaughter commenced. They soon collected a pile of birds, & hats full of eggs."[7]

When McCormick approached in a second boat to join in the fun, he was told to row around the island and catch fish instead. McCormick and his sailors did indeed catch a great deal of fish (for which they vied with many large sharks), which everyone enjoyed after days of preserved rations. Whether that mollified the proud McCormick is extremely unlikely.

As the ship worked its way south toward Cape Horn, Darwin had plenty of opportunity for further explorations on land, and new experiences piled one on top of another. Boxes of specimens were shipped back to Henslow especially because, as Darwin admitted, his collecting of plants was very unselective. The boxes got bigger still when, during two trips to Punta Alta on the Brazilian coast, he found major outcrops of fossils with large mammal bones, both near the shore and in cliffs ten miles inland. At that time there was only one South America fossil in all of England; here was a cornucopia. Among the fossils were what he thought was a rhinoceros, an armadillo, a giant rodent, the sloth *Megatherium,* horse teeth, and a *Mastodon.* Later on he discovered the bones of what seemed to be a giant camel. All these were extinct forms and much larger species than their close living relatives. This was the more exciting because they occurred alongside modern shells. Something, very recently, had caused a major shift in the fauna of these large mammals. What caused them to die out? It seemed to be a confirmation of "the remarkable law so often insisted upon by Mr Lyell, namely, that the 'longevity of the species in the mammalia, is upon the whole inferior to that of the testacea.' "[8]

An early turning point had been Darwin's reading of Lyell's first volume of *Principles of Geology;* then in November 1832, he received from England the second volume. Darwin was well accustomed not only to finding useful information and ideas in books, but also to reading them critically. Lyell was no exception. Darwin read his second volume with an eye both to the places where his own rapidly deepening experiences matched with Lyell's conclusions and to those where they did not.

Given Lyell's intention to make geology a full scientific subject that embraced, drew from, and then explained everything from mineralogy to zoology, botany to geography, it perhaps did not surprise Darwin to discover that the second volume of Lyell's great work scarcely dealt with geology at all. It began with a critical analysis of Lamarck's theories concerning species. Having disposed of Lamarck's ideas about progressive development in the fossil record in his first volume, now Lyell proceeded to a treatment—nothing less than a demolition—of Lamarck's theory of transmutation of species.

Lyell was not only anxious to demolish Lamarck because his theory was tied to the concept of successive development in the fossil record; Lyell was also opposed on religious grounds to the whole concept of the transmutation of species. It will not have escaped Darwin's eye that Lyell's otherwise impeccable logic here deserted him. Lyell set out immutability of species as a premise rather than a conclusion: "The stability of species may be taken as absolute, if we do not extend our views. . . . let a sufficient number of centuries pass, to allow of important revolutions in climate, physical geography, and other circumstances, and the characters (say the transmutationists) of the descendents of common parents may deviate indefinitely from the original type. Now, if these doctrines be tenable, we are at once presented with a principle of incessant change in the organic world."[9] Fixity of species was an important element of the lawfulness of his geological system, the atoms of his geological physics.

Humboldt had expressed the same, eminently practical sentiment: "The natural sciences are connected together by the same ties that link all natural phenomena together. The classification of species, which we should consider fundamental to botany . . . is to plant geography what descriptive mineralogy is to the rocks that form the outer crust of the earth. To understand the laws observed in the rocks . . . a geologist should know the simple fossils that make up the mass of mountains.

The same goes for the natural history that deals with how plants are related to each other, and with the soil and air."[10] For Lyell, Humboldt, and many of their contemporaries, if species were not fixed, all nature was a chaos, never amenable to scientific analysis.

Lyell's first argument against Lamarck was that no one had ever demonstrated transmutation—the evidence was lacking for "one complete step in the process of transformation, such as the appearance, in individuals descending from a common stock, of a sense or organ entirely new, and a complete disappearance of some other employed by their progenitors" even if enough time were allowed.[11] Lyell granted, however, that variation among the individuals of a species leads to the formation of races, but he insisted that nothing further was possible.

Against Lamarck, Lyell pointed out that all breeds of dogs, although superficially very different, are the same species, interbreeding freely. The dogs, cats, ibis, and crocodiles that had been embalmed by the ancient Egyptians are the same species as live today. European domestic animals now living ferally in the Americas have retained their characteristics. But that forced him to dismiss what under other circumstances might have been heuristically valuable—examples like the similarity between a dog and a wolf. "Lamarck has thrown out as a conjecture, that the wolf may have been the original of the dog, but he has adduced no data to bear out such an hypothesis."[12]

Another point against Lamarck was that if there was an inbuilt drive within lineages toward increased complexity, why were there still so many very primitive *living* kinds of animals and plants? (Lamarck had already solved that problem by positing that there had been multiple origins of lineages over time.)

The volume continued with an extensive summary of the known facts and principles in the ecology and distribution of animals and plants. Lyell looked for principles that would account for the widely different floras and faunas existing in different parts of the world and conducted interesting "thought experiments" about what would happen if, for example, the faunas of one region were transplanted wholesale to another. He discussed the variability of animals and plants under domestication, hybridization, and the capacity of all species for overproduction of offspring. This then led him into a discussion of species in relation to their environmental requirements and their capacity to migrate across large distances, even across oceans. All this pointed to the conclusion that creation had not been a single event at a single place, but had occurred in numbers of regional centers or foci.

Among the principles that Lyell inferred from all of this was that "geological monuments alone are capable of leading us on to the discovery of ulterior truths. . . . From such data we may be enabled to infer whether species have been called into existence in succession or all at one period; whether singly, or whether by groups simultaneously; whether the antiquity of man may be as high as that of any of the inferior beings which now share the planet with him, or whether the human species is one of the most recent of the whole."[13]

With his two volumes, Lyell had given Darwin a solid scientific framework for his analyses of South American geology and natural history. Ironically, by laying out a complete argument against transmutation of species, he had also set out the basis of a program of inquiry by which one might test whether the idea might be true. For Erasmus Darwin's grandson, it was almost a challenge. Thomas Henry Huxley later observed: "I cannot but believe that Lyell, for others, as for myself, was the chief agent in smoothing the road for Darwin. For consistent uniformitarianism postulates Evolution as much in the organic as in the inorganic world. The origin of a new species by other than ordinary agencies would be a vastly greater 'catastrophe' than any of those which Lyell successfully eliminated from sober geological speculation. In the end, almost nothing in structure of Darwin's arguments *for* evolution strays from the pattern set by Lyell in his catechism *against* it. A great deal was missing before there would be a scientific theory of evolution. That Darwin provided himself."[14]

In one matter we know that Darwin did in fact read, and quickly react to, Lyell's text. The subject was coral islands. It would be another three years before Darwin saw his first reefs and atolls. But in the intervening period, he conceived of a rival theory to the one that Lyell laid out in this new volume of *Principles,* one that eventually supplanted it. Here the second volume of *Principles* had indeed stimulated Darwin to think deeply about a subject, question the master, and produce his own explanation of the phenomenon.

In mid-December 1832, the *Beagle* reached the mountainous and densely wooded islands of Tierra del Fuego. "To me it is delightful being at anchor in so wild a country." Parties of curious and sometimes hostile Fuegians gathered to watch their progress and were just as carefully observed by the British party. "If their dress & appearance is miserable, their manner of living is still more so. . . . They seem to have no property excepting bows & arrows & spears. Their present residence is under a few bushes by a ledge of rock. . . . It was without exception the most curious & interesting spectacle I ever beheld. I would not have believed how entire the difference between savage & civilized man is. It is greater than between a wild & domesticated animal, in as much as in man there is greater power of improvement." They were essentially naked, living on "limpets & muscles, together with seals and a few birds."[15]

Darwin later wrote that "nothing . . . more completely astonished me, than the first sight of a Savage; It was a naked Fuegian his long hair blowing about, his face besmeared with paint. There is in their countenances, an expression, which I believe to those who have seen it, must be inconceivably wild. Standing on a rock he uttered tones & made gesticulations than which, crys of domestic animals are far more intelligible."[16]

The discomfort of the three Fuegians on board the *Beagle*—Fuegia Basket, Jemmy Button, and York Minster—added to the oddity of the picture.[17] They could speak a reasonable, if halting, English and they wore Western clothes (Fuegia preferred trousers to skirts). The native Fuegians were, by contrast, essentially a Stone Age people ("canoe Indians," as compared with the "horse Indians" of the adjacent mainland, who had considerably more contact with Westerners). Once again, Darwin was faced with the extraordinary variety that could appear within a single species—in this case, his own. Had the Fuegians arisen through "degeneration" from a South American Indian stock, or might they represent an ancient, unimproved version of the human species? It was a sobering experience.

Unfortunately, FitzRoy's hostages had been taken from two different parts of the archipelago and were from different tribes. It made no sense to deposit them all in one place to try to make a stable settlement, but that is what he did, partly because gales prevented them from getting westward into country where York Minster came from. At a place called Woollya, the sailors built three houses, dug gardens, and planted vegetables. Large numbers of Fuegians gathered, including Jemmy's relatives. "At one time there were about 120 [Fuegians]. . . . the men sat all day watching our proceedings & the poor women working themselves like slaves for their subsistence. . . . They asked for everything they saw & stole what they could. . . . On the 27th . . . suddenly every woman and child & nearly every man removed themselves & we were watched from a neighbouring hill . . . neither Jemmy nor York knew what it meant."[18] Despite many misgivings, FitzRoy left them there with their missionary advisor, the Reverend Matthews, and set off further west in the Straits of Magellan.

Returning to the settlement at Woollya a month later, "Matthews gave so bad an account of the conduct of the Fuegians that the Captain advised him to return to the ship. "From the moment of our leaving, a regular system of plunder commenced, in which not only Matthews, but York & Jemmy suffered. . . . it was quite melancholy leaving our Fuegians amongst their barbarous countrymen."[19]

Abandoning the Fuegians to their own devices, at least for the present, FitzRoy set off for a visit to the Falkland Islands. They arrived to find, "to our astonishment, that England had taken possession of the Falkland Islands & that the Flag was now flying."[20] This was Darwin's first experience of oceanic island life since leaving the Canaries. Given the role that evolution on islands would eventually play in the development of his theory of natural selection, it is surprising that Darwin found very little of interest in this new wet and bleak landscape. The remarkable seabird colonies of the Falklands went unnoticed in his diary. By now, he was thoroughly tired of shooting birds and left that work to Syms Covington (originally fiddler and boy to the poop cabin), who had been co-opted as Darwin's manservant. Darwin taught him not only to shoot but also to put up specimens as skins.

⟜ Darwin was not particularly impressed, either, by the fact that the two main islands—East Falkland and West Falkland—had different versions of the native "wolf-fox" (now extinct), the only indigenous mammal. In the light of Lyell's book, these two foxes might have prompted several questions. Were they separate species or two variants of the same species? Had these foxes been specially created on the two islands? Or were they descended from a mainland strain that found its way to one of the islands and then differentiated into two under local conditions (climate, food, vegetation, competitors, predators, and the like)? In his *Narrative,* FitzRoy paid a good deal of attention to these foxes, concluding that the only difference was a "darker and rather thicker furry coat" on the East Falkland form that "may be attributed to the influence of a somewhat colder climate."[21]

For the rest of the year, the *Beagle* worked out of Montevideo, and Darwin had a period of some seven months to escape the ship in journey after journey, exploring inland on horseback with local guides. It was a time for collecting in earnest, but he was already tired of the voyage and the cold, gray surroundings. He was anxious to move beyond the ceaseless back-and-forth of the *Beagle's* surveying work and the prospect of another Antarctic summer (climatically, this was like having four winters in a row). In May 1833, he wrote to his sister Catherine, "I most devoutly trust that next summer (your winter) will be the last on the side of the Horn: for I am become thoroughly tired of these countries: a live Megatherium would hardly support my patience."[22] He took up the same theme two months later: "I am ready to bound for joy at the thoughts of leaving this stupid, unpicturesque side of America. When Tierra del F. is over, it will all be Holidays. And then the very thoughts of the fine corals, the warm glowing weather, the blue sky of the Tropics is enough to make one wild with delight."[23]

In his May letter to Catherine, he also asked for books: "Flemings philosophy of Zoology & Pennants Quadrupeds . . . Davy's consolation in travel, Scoresby's Arctic regions, Playfairs Hutton, theory of the earth, Burchells travels, Paul Scrope on volcanoes, a pamphlet by J. Dalyell on the Planariae, Edinburgh, Caldcleugh travels in S America.—If any of these books are expensive, strike them out." Some of these, like the Scoresby, were books that he had read in Edinburgh; all show the wide extent of his past and current reading.

At the beginning of December, the *Beagle* headed south for what Darwin hoped would be the last time. They were accompanied by the schooner *Adventure* that FitzRoy had taken upon himself to purchase to pursue (extremely effectively) the tricky inshore surveying work. Returning to Tierra del Fuego, they continued their surveys, slowly moving westward until, a little over a year after they had left, they came again to Woollya. "We could see no signs of our friends & we were the more alarmed as the Fuegians made signs of fighting with their bows and arrows. Shortly afterwards a canoe was seen coming with a flag hanging up. . . . we could not recognize poor Jemmy. . . . We found him a naked thin squalid savage. . . . Now he had nothing excepting a bit of blanket round his waist."[24] The puny efforts at setting up the Fuegians as a missionary station had completely imploded. York Minster and Fuegia Basket had taken off for their own country, stealing everything they could find. There was little that the *Beagle*'s people could do for Jemmy except give him some presents.

The ship set off back eastward toward the Falkland Islands, where they discovered a tale of "cold-blooded murder, robbery, [and] plunder."[25] The local "gauchos" had killed most of the settlers. The *Beagle* left with their ringleader and several prisoners on board in chains.

Geology was to be the source of major arguments between Darwin and FitzRoy in the years after the voyage ended, but at this stage it was something they could enjoy together. After the long months of routine survey work in the south, FitzRoy decided that spirits (his and Darwin's) would be lifted by an expedition up the Rio Santa Cruz in southern Argentina. It would be a treat after months of difficult work in conditions that were as cold and gloomy as they were dangerous. Previous explorers had suggested that the headwaters of the Santa Cruz were in the Andes themselves. Captain Pringle Stokes had tried to ascend the river during the *Beagle's* 1826–30 voyage but managed to get only thirty miles inland.

The river mouth was a place where the *Beagle* could be beached for repairs to its keel. Meanwhile, FitzRoy and Darwin, together with a party of twenty-five sailors and three boats, set off upstream. They succeeded in getting about 140 miles inland and to 20 miles short of the mountains, first in the boats and then on foot. But lack of supplies forced them to turn back. With the Andes looming up before them, they returned to the sea, noting that the river had every appearance of having arisen in a lake in the base of the mountains—as indeed we now know it does.

The expedition was not a great success. The *Beagle's* officers and men detailed to the trip found the whole affair "much hard work, & much time lost & scarcely anything seen or gained."[26] But the trip was a geological field course for both Darwin and FitzRoy, "as it was a transverse section of the great Patagonian formation."[27] Working their way inland, the members of the party went through great, flat lava beds and a coastal plain of smooth pebbles with oyster shells. Darwin had already observed that the whole eastern margin of South America showed evidence of processes of pronounced uplift. These plains were in fact the old seabed, and the stones had been rounded by being pounded in a shallow sea.

FitzRoy, who, like Darwin, had been reading Lyell, later described "crossing vast plains composed of rolled stones bedded in diluvial detritus some hundred feet in depth." At the time, he remarked to Darwin, "This could never have been effected by a forty days' flood." We know about this remark because FitzRoy later totally renounced it. Sometime after the voyage ended, a storm erupted as FitzRoy adopted a fundamentalist Christian view of the literal truth of the Bible. Appended to the second volume of the official *Narrative,* which was FitzRoy's narrative of the second voyage, were two essays by FitzRoy. One is a discourse of human migrations through history; the other is entitled "A Few Observations with Reference to the Deluge," and it was intended to counter Lyell's and Darwin's new, nonreligious geology.

FitzRoy recalled his remark to Darwin as having been "an expression plainly indicative of . . . ignorance of Scripture. I was quite willing to disbelieve what I thought to be the Mosaic account." At the time of expedition, he admitted, the two men had happily agreed with "those geologists who contradict, by implication, if not in plain terms, the authenticity of the Scriptures."[28]

On June 10, 1834, the *Beagle* returned to the Straits of Magellan and, after being attacked by bands of Fuegians, left Tierra del Fuego for the third and last time. One of the goals of both the first and second *Beagle* voyages had been to explore and survey routes westward out of the Straits of Magellan. One such passage they discovered was the Magdalen Channel, and FitzRoy took that route out to the Pacific. They did not visit Woollya to check on Jemmy Button.

As they turned to cruise north along the Chilean coast, Darwin saw the Andes close up for the first time. By now, Darwin's conversion from natural history collector to geologist was complete. His letters home to Henslow, his principal scientific correspondent, mark the transition precisely. Letters to Henslow up to July 1834 are largely concerned with collecting. From roughly October of that year, geology occupies at least half of his letters home. After April 1835, except for the usual personal notes, his (few) letters to Henslow are all geological. References in letters to his family related to a projected career as a clergyman show the same trend. His last positive reference to such a career was in May 1833.[29]

Darwin did not stop his collecting. On Chiloe Island, there was yet another island fox "species." As the *Beagle*'s men continued with their surveying and mapping, Darwin once again was free to make expeditions inland. He delighted in the mountain environment, but at this point both Darwin and FitzRoy became ill. Darwin's stomach problems flared up, and he became very weak. He had been staying at some gold mines near Valparaiso and became even more exhausted on the long journey back to the *Beagle* where Bynoe could treat him. "With a good deal of Calomel & rest," he was slowly restored to health. Darwin thought that the problem had been caused by "some Chichi a very weak, sour new made wine, this half poisoned me."[30] Current studies suggest that the real cause might have been the onset of Crohn's disease (see chapter 18).

FitzRoy generously delayed sailing in order to give more time for Darwin to recover, but he was undergoing his own crisis, putting the entire expedition at risk. At Valparaiso he had a complete nervous breakdown. Dreadfully overworked and burdened with all the responsibilities of the voyage, throughout the expedition he had consistently exceeded, or even disobeyed, Admiralty instructions. As Captain King had discovered during the first South American voyage, square-rigged ships like the *Beagle* were inherently unsuitable for the sort of inshore work that detailed mapping required. In those rock-strewn, stormy waters, a small schooner (like the sealers that plied those waters) was necessary. FitzRoy saw this need and had hired or bought smaller schooners whenever they were needed. Although FitzRoy's results totally vindicated his actions, he was forced to make the commitments first and ask for permission later.

The responses from the Admiralty necessarily arrived long after the fact, when the expenditures were made and the work was done. They consistently refused FitzRoy's requests for compensation; the latest case was the *Adventure.* At Valparaiso, FitzRoy received the last of these letters. In having to make good the various hires and purchases of small vessels, FitzRoy was almost ruined financially, and the rejection of his methods by the Admiralty caused him equal distress.

FitzRoy wrote to Beaufort on September 26, "Troubles and difficulties harass and oppress me so much that I find it impossible either to say or do what I wish."[31] At the end of October he gave up. Darwin wrote to Henslow, "The selling of the schooner & its consequences were very vexatious: the cold manner the Admiralty (solely I believe because he is a Tory) have treated him and a thousand other &c, &cs has made him very thin and unwell. This was accompanied by a morbid repression of spirits, & a loss of all decision & resolution. The Captain was afraid that his mind was becoming deranged (being aware of his hereditary disposition)."[32]

In a deep, dark depression, FitzRoy could no longer face going back to finish the work around Cape Horn. He decided to quit the ship and return home, leaving Wickham, the first lieutenant, in charge of the *Beagle* with instructions to complete the Cape Horn survey and then take the ship back to England. Darwin and the other officers, however, managed to persuade FitzRoy that he had more than done his duty with respect to South America and that he could honorably continue the voyage west, across the Pacific. Apart from his pleasure at the restored mental health of his friend and colleague, Darwin was delighted that the ship would not return home before completing the leg across the Pacific, and thus the circumnavigation. "Hurra Hurra it is fixed the *Beagle* shall not go one mile south of C. Tres Montes . . . & will be finished in about 5 months."[33]

In the end, it was another eight months before the *Beagle* finally left South America, and in the meantime there were new adventures. His health recovered, Darwin set off inland for more explorations. On February 20, 1835, the ship was at Valdivia, and Darwin was ashore riding when there was a massive earthquake. "I was on shore & lying down in the wood to rest myself. It came on suddenly & lasted two minutes (but appeared much longer)."[34] At first the quake seemed relatively minor; no buildings in the town were actually destroyed. Four days later they moved on to Concepción, which had been much harder hit by the shock and after-shocks. Even further along the coast, a tsunami wave had done a huge amount of damage. "All the towns . . . Concepción and St Jago have been destroyed."[35] In places the land had been raised up by eight or nine feet. Here was dramatic and conclusive proof of elevation of the land by earthquakes, just as Lyell had proposed, and confirmed by professional surveyors.

With his confidence recovered, FitzRoy actually hired yet another auxiliary vessel, the schooner *Constitución*. Darwin continued his inland explorations, engrossed by the geology as well as the natural history. On rejoining the *Beagle* in July 1835, he discovered that he had missed the excitement caused by the wreck of HMS *Challenger* at

Arauco, south of Concepción, almost certainly because of subsurface changes caused by the earthquake. When news of the *Challenger*'s sinking reached England, Darwin's family was particularly anxious because he had just written to say that he was dispatching the next installments of his journal home with it. Luckily, they were put onto a different ship and reached England safely. What they did not yet know was FitzRoy's dramatic role in rescuing the *Challenger*'s men.

In this episode we see the dynamic, even frantic, side of FitzRoy's bipolar personality. Frustrated by the reluctance of the local naval commander (Commodore Mason, flagship HMS *Blonde*) to take seriously the news of the wreck of the ship (on which FitzRoy had once served), he set off down the coast on horseback, on a wild sixty-mile ride, to find where the survivors had come ashore. Having done that, and seen that they had erected defenses against the surrounding hostile locals, he rode pell-mell back to Concepción and bullied Mason into setting off to the rescue, acting as the pilot. It was vintage FitzRoy at his dashing best.

On September 7, 1835, the *Beagle* finally headed out west across the Pacific, starting the last half of its long voyage. The crew's first landfall, fatefully, would be the Galapagos Islands, a set of volcanic islands some six hundred miles from the coast. The whole archipelago is of volcanic origin; some volcanoes were still active, and in many places the lava flows looked as though they had just congealed. The islands lie on the equator, although the climate is cool because of the Humboldt current that bathes them, and they display a wide range of climate, soil, and vegetation conditions. Two-thirds of the resident birds, and all the reptiles, are unique to the islands. A striking feature of the wildlife was (and still is) its tameness. "The birds are Strangers to Man & think him as innocent as their countrymen the huge Tortoises. Little birds, within 3 or 4 feet, quietly hopped about the Bushes & were not frightened by stones being thrown at them. . . . I pushed off a branch with the end of my gun a large Hawk."[36]

In the five weeks they spent at the islands, Darwin's collecting (mostly done by his servant, Covington) was scrappy, and the labeling of specimens from the different islands was confused. Probably it didn't seem to matter; the islands were so close to each other and the birds all so similar. Over dinner one day, Darwin was told by the deputy governor of the remarkable case of the land tortoises—they were so distinctly different, island by island, that (he claimed, exaggeratedly) anyone could tell where a specimen came from just by looking at the shape of the shell.

As for the origins of this curious fauna and flora, Darwin assumed that the birds and reptiles of different islands were simply variants of mainland species. As in the case of the foxes of the Falklands and Chiloe, and as Henslow had seen in his 1830 experiments with plants, local variation was intimately tied to differences in local conditions.

By now—luckily, as it happened—several of the officers were making their own natural history collections, FitzRoy among them. It was not until long after the *Beagle* had left the Galapagos that Darwin started to think seriously about the significance of the unique faunas and floras of the islands. The scope of those collections then became important.

After the Galapagos, the *Beagle* made landfalls in Tahiti and New Zealand before making a lengthy stay in Australia. There Darwin was particularly struck by the fact that the continent, like South America, had its own unique fauna and flora, particularly the large mammals, not one of which was held in common with another part of the world. Here was surely an example of a separate "centre of creation."

As the ship made its way westward across the Indian Ocean, one more special point of the voyage was reached. At the Cocos Keeling Islands, Darwin was able to test for himself whether the theory of coral reef formation that he had been developing was correct. Coral islands come in various configurations, the most dramatic being the complete ring with an inner lagoon. Any theory of their formation had to take account of the fact that the coral organisms grow only where there is light. Lyell's view was that coral atolls formed on the rims of extinct, slightly submerged, volcanoes—hence the ring shape. Darwin, having read Lyell's explanation, saw that a more likely explanation was that the coral formed first around the margin of an emergent peak. Then the peak became submerged and, as it did so, the coral steadily grew upward in response until there remained only the ring itself. If Lyell's theory was right, the coral would be only a few meters deep. If Darwin was right, the living coral on the surface would be supported on an accumulation of dead coral and could be extremely deep, depending on how long the process had continued. FitzRoy took soundings and brought up dead coral from the depths: Darwin was right.

After what was beginning to feel like an eternity, finally the *Beagle* reached the Cape of Good Hope and after a stop there, where the expedition members visited none other than Sir John Herschel, who was there making astronomical observations, the long journey northward—homeward—began.

On the way back up the South Atlantic, Darwin worked on organizing the notes and information that went with the vast array of specimens he had collected during the voyage. He started making entries in notebooks, each devoted to a different subject: geology, birds, shells, fish, and so on. In this process Darwin pulled together his observations on the distributions of organisms. They showed that related species tended to replace each other geographically. He found a particularly good example of this relationship in the two kinds of rheas (ostrich relatives) he had collected in South America. One (the common rhea, or *Avustrez*), was more northerly in distribution and larger than the rarer, more southerly, one (the *Avustrez petise*). Darwin had not known there were two species until the Indians told him. His first encounter with the *Avustrez petise* was with one that Conrad Marten (the artist who replaced Augustus Earle) shot at Rio Santa Cruz. Unfortunately, they ate it, Darwin having realized only at the last moment what it was. "Fortunately the Head & neck had been preserved." He noted that the common rhea lived in Patagonia "as far south as a little south of the R. Negro," while "the Petise takes its place in Southern Patagonia, the part about the R. Negro being neutral territory."[37]

Then Darwin came to work up the birds from the Galapagos Islands in his Ornithological Notes. Of the Galapagos mockingbirds, he wrote: "Thenca (Galapagos mocking bird) I have specimens from four of the larger islands. . . . In each Is. each kind is *exclusively* found: habits of all are indistinguishable."[38] He assumed they were all minor variants, rather than distinct species. But he also immediately started to make associations and to wonder whether these "varieties" had some greater significance.

"When I recollect, the fact that . . . the Spaniards can at once pronounce, from which Island any tortoise may have been brought;— when I see these Islands in sight of each other, & possessed by but a scanty stock of animals, tenanted by these birds, but slightly differing in structure & filling the same place in Nature, I must suspect they are only varieties. The only fact of a similar kind of which I am aware, is the constant asserted difference—between the wolf-like Fox of East & West Falkland Islds.—If there is the slightest foundation for these remarks the Zoology of Archipelagoes—will be well worth examining: for such facts [*would* added] undermine the stability of Species."[39]

Regarding the Galapagos finches, he was less sure on the issue of species versus variants. He wrote: "Here these Finches are in number of species & individuals far preponderant over every other family of birds. Amongst the *species* [emphasis added] of this family there reigns (to me) an inexplicable confusion. Of each kind, some are jet black, & from this, by intermediate stages, to brown; but my series of specimens would go to show, . . . that color is proper to the old cock birds alone. . . . Moreover . . . A gradation in the form of the bill, appears to me to exist.—There is no possibility of distinguishing these species by their habits, as they are all similar, & feed together (also with doves) in large irregular flocks."[40]

Darwin repeated his conclusions when he later wrote up his *Journal and Remarks* (*Voyage of the Beagle*): "The most curious fact is the perfect gradation in the size of the beaks in the different species of Geospiza, from one as large as that of a hawfinch to that of a chaffinch, and . . . even to that of a warbler. . . . I very much suspect, that certain members of the series are confined to different islands; therefore, if the collection had been made on any *one* island, it would not have presented so perfect a gradation. It is clear, that if several islands have each their peculiar species of the same genera, when these are placed together, they will have a wide range of character. But there is not space in this work, to enter on this curious subject."[41]

FitzRoy wrote about the Galapagos birds in his own *Narrative* of the voyage, and he emphasized that their different beaks seemed to be adapted for different conditions and purposes: "All the small birds that live on these lava-covered islands have short beaks, very thick at the base, like that of a bullfinch. This appears to be one of those admirable provisions of Infinite Wisdom by which each created thing is adapted to the place for which it was intended. In picking up insects, or seeds which lie on hard iron-like lava, the superiority of such beaks over delicate ones, cannot, I think, be doubted."[42]

At the time, Darwin seems to have thought there were at least four groups of species within the finches of the Galapagos, and he referred to them in terms of analogy with grosbeaks and orioles as well as finches. The "inexplicable confusion" stemmed in part from the fact that Darwin could not sort them either by color or bill shape, understandably not realizing that similar variations in both features existed in separate species groups—the cactus finches, ground finches, and so on—and on different islands.

Another part of Darwin's confusion over the kinds of finches on the Galapagos arose because they had not been collected with a careful eye to which island each bird came from. This is understandable, given the apparent homogeneity of the birds. By the time Darwin came to work up his Ornithological Notes, it was almost too late. At least he had recognized something of the status of the mockingbirds and, luckily, FitzRoy and other officers later turned out to have been more punctilious. Eventually, the pattern of finch diversity on the islands would be worked out by experts in London.

The Ornithological Notes make it quite clear that at this time (toward the very end of the voyage), Darwin was still fully committed to the notion that most if not all cases of closely similar animals and plants occupying adjacent, and similar, environments were examples of intraspecific variation. In that case, what did he mean by the phrase "[would] undermine the stability of Species" in the entry quoted above? He might have written the words approvingly, or it is equally possible that he used the phrase dismissively. In either case, it indicates Darwin's full familiarity with the concept of transmutation and perhaps his first new thoughts on the subject.

PART THREE

Natural Selection

First Thoughts on Evolution

On October 2, 1836, Darwin arrived in England full of both excitement and anxiety. All thoughts of a quiet career in a country parsonage were now forgotten. He had a new life to start, both a personal life and a scholarly one, and his entry into scientific society turned out to be easier than he had feared. The way had been prepared for him wonderfully. Because Henslow had published some of Darwin's letters and had sent various of his more exciting specimens (like the Punta Alta fossils) to experts to examine, Darwin was already quite well known by reputation in both Cambridge and London. Not only was he eager to meet the leaders of Britain's natural science, they were also keen to meet him.

After spending the minimum respectable time in Shrewsbury with his family, Darwin hurried to London, where he stayed with his brother Erasmus. At the top of Darwin's list of men to cultivate was Charles Lyell and, less than a month after his arrival in England, he was able to report his successes to Henslow: "Mr Lyell has entered in the *most* goodnatured manner, & almost without being asked, into all my plans."[1] (Given the low esteem in which Lyell's geology was held in Cambridge, this enthusiasm may have been less than tactful.)

Darwin dined as Lyell's guest at the Zoological Society and Geological Society. Lyell proposed him for membership in the Geological Society and on January 4, Darwin read there a paper entitled "Observations of Proofs of Recent Elevation on the Coast of Chili"—something that would have been impossible without Lyell's imprimatur. (He noted to Fox, ironically: "I was proposed to be a F.G.S. [fellow of the Geological Society] last Tuesday. It is, however a great pity that these & the other letters, especially F.R.S. [fellow of the Royal Society] are so very expensive.")[2] Whatever his worries had been about being out of circulation for so long, and having failed yet to embark upon a career, they were unnecessary—he was already accepted as a (young) peer among the scientific establishment.

Darwin's first thought had been to settle in Cambridge to start his new life. He would be comfortable there among supportive scholarly friends. After staying a few days with the Henslows, he set up residence on FitzWilliam Street. But Darwin quickly found the social life at Cambridge was too demanding on his time, and the truth was that the center of his scientific world had shifted. Within weeks, it was clear that meetings of the Geological Society and Royal Society were going to be more stimulating than Henslow's evening soirees, and far more useful to his career goals. Sooner or later he would have to move to London.

Significantly also, Erasmus was living in London and would be available, once again, as a mentor and support. In March, Darwin took rooms on Great Marlborough Street, not far from his brother. Erasmus greeted him with open arms, folding him into his social network of literary types like Thomas Carlyle and reformers such as Harriet Martineau.[3]

Ever serious and focused on his work plans, Darwin discovered that London also had its drawbacks. His new scientific friends in London were far too quarrelsome for his taste. At an evening meeting of the Zoological Society, "the speakers were snarling at each other, in a manner anything but like that of gentlemen."[4] Overall, Cambridge had been more civilized, and Darwin was a man who had great disdain for what he thought was unseemly behavior.

Mainly, he was far too busy, being pulled in different directions by his many interests, and he missed the pleasures of the countryside. He wrote to Leonard Jenyns, "I thought Cambridge a bad place from good dinners & other interruptions, but I find London no better, & I fear it may grow worse.—I have a capital friend in Lyell, and see a great deal of him. . . . I miss a walk in the country very much; this London is a vile smoky place, where a man loses a great part of the best enjoyments of life. But I see no chance of escaping even for a week from this prison, for a long time to come."[5]

Moving to London allowed Darwin to develop the kind of close relationship with Charles Lyell that he needed. Lyell managed Darwin's scientific life much as Erasmus did his social life. In addition to the Geological Society, Lyell got him elected to the Athenaeum, which the shy Darwin "fully expected to detest."[6] But the first time he dined there—with the geologist William Henry Fitton, the microscopist Robert Brown, the botanist Francis Boot, and William Sharp MacLeay—he found it very valuable. (Darwin was fortunate in having returned to England just before MacLeay (author of the highly influential Quinary System of classification) emigrated to Australia.)

Lyell was thrilled with his new colleague, someone who so perfectly appreciated his geological ideas and who had such a wealth of firsthand experiences. Lyell's geology was still controversial and Darwin, having taken Lyell's systems into the field and validated them, was one of the first real fruits of his geological teachings. The fact that Darwin had already modified many of Lyell's points, particularly on the subject of coral reef formation, seemed not to disturb Lyell at all. He welcomed the challenge and saw in Darwin a true philosophical geologist after his own heart. Within a couple of months, Lyell wrote to Darwin, "I am just revising what I have said in my Anniversary Address, of you & your new Llama, Armadillos, gigantic rodents, & other glorious additions to the Menagerie of that new continent, which was heaved up. . . . I could think of nothing for days after your lesson on coral reefs, but of the tops of submerged continents." The letter continued, very presciently, "But do not flatter yourself that you will be believed, till you are growing bald, like me with hard work, & vexation at the incredulity of the world."[7]

The immediate consequence of all this recognition was that Darwin quickly became swamped with work. No sooner had he been made a member of the Geological Society than he was asked to join its council, and then to become its secretary. He declined the secretaryship, but felt he had no choice but to accept when it was offered again the following year.

In those first six months after the *Beagle* voyage, Darwin's self-imposed workload was prodigious. Not only had he important ideas to develop and circulate, he had discovered a love of writing. His first project was to complete the development of his *Beagle* diaries into the book—*Journal and Remarks, 1832–1836*—that later became the immensely influential *The Voyage of the Beagle*. At the same time, he set to work on a number of scientific papers in geology and began to plan his book on coral reefs.

Another pressing major task was to arrange for study of his collections and for the publication of the scientific results—at the government's expense. He had foreseen that to get the most out of his collections he would need to farm them out to experts willing to analyze them—and quickly (rather than have them sit around in someone's cabinet drawers). Finding these people was a chore, but it brought him to meet all the important natural scientists of the day. By late spring, results of all this work were coming in: Richard Owen (Britain's leading anatomist and paleontologist, based at the Royal College of Surgeons) was sending information about the fossil mammals; John Gould (artist and ornithologist) was reporting on the birds; Thomas Bell at University College London wrote up the reptiles; Leonard Jenyns reviewed the fish and George Waterhouse the mammals. Darwin wrote a geological introduction and added many notes to the bird and mammal section. Grant asked to be allowed to work up the corals, but Darwin refused him—the wounds had not healed there. Darwin met with the chancellor of the Exchequer in August and secured the funding. *The Zoology of the Voyage of H. M. S. Beagle* was published in five volumes between 1839 and 1843.[8]

As for relations with his old shipmate, friend, and coadventurer Robert FitzRoy, things cooled very quickly. After the voyage they saw rather little of each other—there were a few formally necessary afternoon teas and a dinner or two with FitzRoy and his new wife. Soon there was a major falling-out. When FitzRoy saw the draft of Darwin's *Journal and Remarks,* he exploded. The problem was just not the science, but the dedication. After five years during which he and his officers had smoothed the way for Darwin in every possible sense (except

for the ocean rollers themselves), they were not adequately acknowledged. Darwin had not properly thanked Bynoe, who had probably
saved his life at least once.

The charitable view is that this was purely a mistake, caused by in-
experience in the matter of authorship, and perhaps by the rush in getting the manuscript ready for the printers while in a whirlwind of
activity in London. Darwin changed the introduction, but still very fee-
bly. In the final version, he wrote: "I hope I may here be permitted to
express my gratitude to [FitzRoy]; and to add that, during the five years
we were together, I received from him the most cordial friendship and
steady assistance. Both to Captain FitzRoy and to all the Officers of the
Beagle, I shall ever feel most thankful for the undeviating kindness with
which I was treated, during our long voyage." An added footnote reads:
"I must likewise take this opportunity of returning my sincere thanks to
Mr. Bynoe, the surgeon of the Beagle, for his very kind attention to me
when I was ill at Valparaiso."[9] This left Darwin open to the charge that
would dog his career: that he was unwilling to give credit to others.[10]

Darwin's *Journal and Remarks* was based on the extensive travel
diaries that he had kept during the voyage, backed up by the mass of
detailed information on animals, plants, and geology that he had collected in his notebooks. It was so beautifully written that anyone could
read it, and yet it contained a mass of detail in natural history. In every
way it was a worthy successor to Humboldt's *Personal Narrative.* Darwin finished the manuscript around July 1837, and by September he
and Henslow were reading proofs. He "sat in silent admiration at the
first page of my own volume." (He also noted that a copy editor had introduced a number of errors.)[11] When finally published in 1839, the
book sold well—far better than the rather stodgy volumes that FitzRoy
had written, much to the latter's chagrin.

Throughout his life, Darwin would usually have many projects, at various stages of maturity, on the boil at the same time. His mind was too
active to proceed otherwise. Trying to compartmentalize his work was
more or less easy when the subjects did not clash. Often they did, and

the effect was always seen in his health, which lurched from good to poor as his anxiety levels rose.

In the spring of 1837, while he was drafting *Journal and Remarks,* worrying about his collections, and recruiting authors for the *Zoology* of the voyage, Darwin was abruptly faced with his first major scientific dilemma since Grant's flustrae affair. He began to think seriously about the issue of transmutation of species (or, as he preferred, "transformism"). By September his heart palpitations had returned: "My doctors urge me *strongly* to knock off all work & go and live in the country for a few weeks."[12]

The critical issue for Darwin's transformism turned out to be the birds and plants of the Galapagos Islands.[13] As the Ornithological Notes show, during the voyage itself, Darwin's working assumption had been that the island forms were only varieties of mainland species. He assumed that very similar parent stocks would be found on the mainland (which he had not visited at that latitude). The same notion applied to the Galapagos tortoise, which he thought was the same species as that found on Aldabrah and other islands in the western Indian Ocean (Darwin assumed in this case that the Galapagos species was the ancestor of the Indian Ocean variant). As we have seen, while the notebook entries show that Darwin was familiar with the concept of transmutation, as was any well-informed naturalist of the day, he had not accepted the idea. He favored the view that the complex patterns of distribution of different species, region by region, were due to their having being formed at a number of separate centers of creation.

Owen had identified Darwin's fossil mammals from Punta Alta as including extinct species of sloth, an armadillo, and something related either to a camel or llama. In March 1837, Darwin learned from Gould that the two South American (ostrichlike) rheas were distinct species (even though the specimen of the southerly species—later named *Rhea darwinii*—was only the remains of the bird Darwin ate for dinner).[14]

By gathering together specimens from various collections (like FitzRoy's) that had been made on board the *Beagle,* Gould showed that the mockingbirds and finches Darwin had collected on the Galapagos were all distinct species, not closely related to birds on the mainland.

There were at least eleven species of finches and, whereas Darwin had thought they belonged to four different groups, Gould showed that there was just one—unique to the archipelago. Moreover, each island had its own subgroups of these finch species.[15]

As the islands of the archipelago were of geologically very recent volcanic origin, there were only two possibilities: either the Galapagos species had been newly made by the Creator especially for those remote islands, or they had originated in situ by diversification from one or more migrant ancestors. It is generally agreed that Gould's decision that the mockingbirds and finches were all distinct species pushed Darwin firmly into the "transformist" camp. The question was: what would he do about it? (Gould seems to have stayed out of the debate.)

Darwin was well aware that it would be stupid even to hint at evidence for transmutation of species in *Journal and Remarks*. He would either be ignored or derided. It was such a huge and contentious issue, with such a long history, that any publication purporting to make the case for the mutability of species would have to be meticulously documented and overpoweringly argued. So, in his manuscript, when he came to describe the fascinating diversity of species, island by island, on the Galapagos, Darwin merely stated that "there is not space in this work, to enter on this curious subject." In the 1845 edition of the *Journal* he still carefully did not show his hand, only dropping what may be seen now as the most gentle of hints. From the unique geology of the islands, he wrote, "we seem to be brought somewhat near to that mystery of mysteries—the first appearance of new beings on this earth."[16]

Twelve months after the voyage ended, Darwin had begun to demonstrate an extraordinary capacity to manage several careers at once. Central to everything was his ability to strike to the cores of multiple ideas, arguments, and data sets to find new truths. Jameson, Hope, Grant, Lamarck, Locke, Paley, Lyell, Henslow, Sedgwick, and many, many others had contributed to his formal learning. The breadth of his reading continued to be formidable. The voyage itself, coupled with many prior years of amateur natural history, had given him a wealth of experiences in the natural world. All this was processed in Darwin's bear trap of a brain—meticulous, questioning, imaginative, persistent, brilliant—always open to new ideas and thoughts, forgetting nothing important, never glossing over a tricky question.

From the late spring of 1837, the idea of transmutation of species started to preoccupy him. And whatever else we know about Darwin, we know that he hated idle speculation and conjecture. He would not have entertained the idea seriously unless he was fairly sure he could make a significant contribution to the subject—that he could find, or had already found, some of the answers. He was brilliant; but even more so, he was careful, and he never wasted time on fruitless fishing expeditions.

Gould's news about the Galapagos species did not suddenly propel Darwin into thinking about transmutation for the very first time. Transmutation was—as a theory—very familiar to him. A great deal of effort has been expended by scholars in trying to determine the *precise date* at which Darwin began the pattern of thinking that led to his theory of transmutation of species by means of natural selection. This is a vain hunt, because there never was such a single, critical point. One could not be a natural scientist without knowing what the existing transmutationist ideas were, whether or not one subscribed to them. Darwin had known about the subject at least since his Edinburgh days; everything he later learned in Cambridge had been reinforced and extended by Lyell in his second volume of *Principles.*

In tracing the tangled course of Darwin's thoughts, therefore, one has to distinguish between general references to transformist concepts that derive from other workers and those that show Darwin developing his own ideas. Most students of Darwin have assumed that he was heading more or less directly toward the subject of transmutation in the spring and summer of 1837. Some have even thought that Darwin intuited his entire theory in spring 1837.[17] A more likely interpretation is that Darwin's aims were at first both broader and less defined.

The pattern of Darwin's changing ideas can be traced reasonably accurately because of his habit of recording thoughts and facts in his scientific notebooks.[18] In them, he made quick (often bafflingly elliptical) observations about his scholarly activities—the books he had read, thoughts quietly explored, questions asked, and authorities sought for.[19]

Like the Ornithological Notes, Darwin's small Red Notebook was started during the last stages of the voyage. Until the end of the voyage, its entries were concerned largely with geological subjects. Darwin began Notebook A, which also mostly concerned geology, in mid-1837, and it ran to December 1839. In January 1837, Darwin started writing in the Red Notebook again, and this time there were more references to natural history. Among the latter are some entries that show him considering transformist views between January and June. Then, in July 1837, he started the first notebook devoted to "transmutation of species." That was Notebook B, which was followed in an accelerating series by Notebooks C (March 1838), D (July 1838), and E (October 1838).

Very many of the notebook entries were written as memoranda to himself—little instructions to look more deeply into something—with a view to inclusion in future publications. It is not possible here to do justice to the immense range of erudition, the huge number of different topics researched, and the tangled networks of speculation that are contained in the notebooks. In the following pages I have necessarily focused on elements most directly involved in the evolution of the idea of natural selection.

Darwin's notebook entries are interesting for the fact that almost everything there is directed toward, or colored by, his *Beagle* experiences. When he muses over extinction, for example, his examples are not ammonites or ichthyosaurs or trilobites—but the sorts of fossils he dug up in South America—the mastodon, *Megatherium*, the horse, and the "camel." The notebooks also show him to be preoccupied with exploring a wide range of facts and concepts in science, and they are therefore interesting for what is *not* included. Above all, reading the notebooks shows the vast range of intellectual debts that Darwin owed to others as he developed his theory—a useful counterpoint to the impressions he gives in the *Autobiography*.

While his eventual masterpiece *On the Origin of Species* (1859) shows that he must have been deeply influenced by his reading and personal experiences in political, social, philosophical, and literary spheres, he did not draw upon these subjects overtly for ideas or metaphors in the very early years. This leaves some intriguing questions. For example, did Darwin's view of evolution as "progressive," producing the unfolding history of life seen in the fossil record, stem only "internally" from his reading of contemporary science, or did it also (perhaps even mainly) reflect the progressivism of the contemporary social and political realm? To what extent were his views on the races of humans dependent on current scholarship, and how did his views (and those of the authors he consulted) reflect the debates over slavery (in which the Wedgwood family was so prominent)? The evidence of the notebooks suggests that at this early stage of his transformist inquiries, Darwin was focused on the science. Only after he was sure that he had the core of an idea did he open notebooks on metaphysical issues, for example.

Darwin was intensely ambitious before he set out to crack the problem of "transformism." While completing the work on *Journal and Remarks,* he had set himself a long-term scientific agenda with the immodestly lofty goal of transforming the philosophical basis of natural science. He set out to use the huge variety of his previous experiences and his vast reading to discover what he termed a "new system of Natural History." Such a system, he believed, would "describe the limits of form" and explain the extraordinary facts of animal and plant diversity—both taxonomic and geographical—that he had observed around the world.[20] Intellectually, this scheme would fall somewhere between the natural history of collectors and taxonomists and the natural philosophy that was normally the province of physicists, mathematicians, and (as Herschel had written) most recently the geologists. Darwin wrote in his Notebook B, with obvious reference to the subtitle of Erasmus Darwin's *Zoonomia: "The Grand Question, which every naturalist ought to have before him, when dissecting a whale, or classifying a mite, a fungus, or an infusorian. Is 'What are the Laws of Life?'"* (emphasis in the original).[21]

It was out of this broader approach that he distilled his ideas on transmutation; within it, evolution became the central principle of all biology. He accomplished his daringly immodest goal.

Nothing Darwin attempted would be (or could have been) done ex nihilo; there were already schemes purporting to represent new systems. Darwin had been educated in them, both formally and informally, since Edinburgh days. The existing models and philosophical elements included Herschel's principles of natural philosophy and vera causa; the principles of natural theology set out by Paley and adopted by all of Darwin's Cambridge teachers; the successive development principle that preoccupied people like Sedgwick; and Lyell's principle of uniformitarianism and geological cyclicity.

One of the most important bodies of work overtly claiming to represent a new era of rational, philosophical, natural history was MacLeay's Quinary System (see chapter 9). In 1819, MacLeay had set out on just such an intellectual journey as Darwin's. His system was central to Darwin's early thinking because it was an attempt to solve the perennial question: what is the deeper significance of the patterns of similarity and difference displayed by diverse groups of organisms? This was precisely the realm of the "philosophical naturalist," whose interests lay between those of the collector and the pure theoretician.

Darwin's predecessors in the field of transmutation of species had each developed their ideas from foundations in different experiences and interests. Georges-Louis Leclerc, comte de Buffon, in Paris had been fascinated by "generation" and control of embryological development, starting with the sperm and egg. From his ideas about the internal control of form, he proposed that the path of embryological development might be affected by the environment: this would at least produce species—negatively—"degeneration." The great apes, in that view, had arisen as faulty versions of humans. Erasmus Darwin started from the same viewpoint but emphasized the effects of environment, use and disuse, and will in creating a potential for positive change. Lamarck, like Erasmus Darwin, was impressed by the progressive nature of the fossil record.[22] MacLeay, on the other hand, while preoccupied by the issue of relationship, similarity and difference, saw the world as entirely static and unchanging.

The notebooks show that, as a serious scholar, Darwin began where all natural history begins: with identifying and classifying species, and he added to that a thorough understanding of the fact that the earth was a constantly changing system. Fresh from his *Beagle* experiences, Darwin brought to the subject his training as a systematist and his knowledge, developed during the voyage, of the bafflingly complex patterns of biogeography. Through all his analyses, Darwin took the fundamental position that the logical ordering of information through systems of classification was the key to any deeper understanding of geographical and geological patterns in nature.[23]

Darwin had started the voyage as a conventional Christian. His notebooks for early 1837 show that at first he followed Lyell and his Cambridge teachers in the hypothesis that new species arose through special creations as a direct result of the exertion of the ultimate power (first cause) at different centers of creation. By this, one could explain the fact that Australia and South America had such different faunas from that of Europe and Africa.[24] As the world's diversity became better documented by people like Linnaeus, Humboldt, and Darwin himself, every time a new pocket of diversity was discovered, a new center had to be postulated. What was far from clear was whether creation had happened in these centers all at the same moment, or whether there had been a succession of centers over time and across space. Had there been a separate (very recent) center just for the Galapagos, for example?

While at the Galapagos, Darwin wrote in his journal, "It will be very interesting to find from future comparison to what district or 'centre of creation' the organized beings of the Archipelago must be attached."[25] Later, in an entry from Australia, Darwin described himself as "lying on a sunny bank and . . . reflecting on the strange character of the animals of this country. . . . An unbeliever in everything beyond his reason might exclaim, 'Surely two distinct Creators must have been at work; their object, however, has been the same & certainly the end in each case is complete.' Whilst thus thinking, I observed the conical pitfall of an Lion-Ant. . . . Without doubt the predaecious Larva belongs to the same genus but to a different species from the European kind.— Now what would the Disbeliever say to this? Would any two workmen ever hit on so beautiful, so simple, & yet so artificial a contrivance. It cannot be thought so. The one hand has surely worked throughout the universe. A Geologist perhaps would suggest that the periods of Creation have been distinct & remote the one from the other; that the Creator rested in his labor."[26] This sentence was repeated in *Journal and Remarks,* in both the 1839 and 1845 editions. In fact, in reading them and Darwin's early notebooks, one is struck by the number of times references to the "centres of creation" concept appear, together with other references to "the Creator" (there is at least one reference in the Ornithological Notes of late 1836). Gould had shattered this whole concept for Darwin.

Another starting point for Darwin was to follow Lyell (and the majority) in believing that species, once created, were not just unchanging but were always discretely different one from another. He rejected Lamarck's view (which originated with Linnaeus), then growing in popularity, that if we were ever to discover all the species of the world, we would find that they graded imperceptibly ("insensibly") one into the other. That, of course, was both a premise and a consequence of Lamarck's transmutationist theory, just as the fixity of species was a premise of the creationist view. On the other hand, as the entries for the Galapagos fauna in the Ornithological Notes show, Darwin had a keen sense for the scale of variation that was to be seen in some species (and not in others).

Extinction was a familiar concept to Darwin, although there were still a few authorities who refused to believe it ever occurred. His early views on extinction were colored by his adoption (or at least serious consideration) of the idea that the Almighty had allotted a particular span of time for each species (analogous to our own individual threescore years and ten), after which they died and were replaced. As he wrote in the *Journal and Remarks,* "All that at present can be said with certainty, is that, as with the individual, so with the species, the hour of life has run its course, and is spent."[27] Darwin was much puzzled about the relationship between adaptation and survival or extinction. Of the animals in his experience, the large mammals in South America (and North America and Australia) had become extinct relatively recently, but there had apparently been no drastic change of climate or other circumstances to account for that. He made a note to himself: "Inculcate well that Horse at least has not perished because too cold:—with discussion of camel urge S. Africa productions."[28] On page 129 of the Red Notebook, in considering the extinction of his fossil "llama" from Punta Alta, he wrote that it "owed its death not to change of circumstances. . . . Tempted to believe animals created for a definite time:—not extinguished by change of circumstances." Later entries show that he also thought that entire groups had a similarly finite lifespan.

Darwin's first notebook entry referring explicitly to transmutation was the entry in the Ornithological Notes, page 74. But the language does not tell us whether he thought that transmutation of species (undermining the stability of species) was intellectually a good thing, or a great and dangerous fallacy. On the whole, the entry seems not to be a statement in favor of transmutation.

Having taken up the Red Notebook for a second time, Darwin shifted gears, as it were, from geology to animals and plants. He started to address the possible causes of biogeographical distributions that would recur through his subsequent thinking (where the reader can decode his eccentric punctuation). He instructed himself to: "Speculate on neutral group of 2. ostriches; bigger one encroaches on smaller.—change not progressif: produced at one blow. If one species altered: Mem: My idea of Volc. Islands. Elevated. then peculiar plants created. if for such mere points; then any mountain, one is falsely less surprised at new creation for large.—Australia = if for volc isld. Then for any spot of land. = Yet new creation affected by Halo of neighbouring continent: as if any creation taking place over certain area must have peculiar character."[29] (Presumably this was written just after Gould had announced the specific status of the rheas.)

In this entry he seemed to be speculating whether, as new volcanic islands are formed, new species are formed for the new circumstances. And, if this were the case for small areas like islands, would it also be true for changes in large areas of land and their species (the latter case being even easier to imagine than new species being formed on some tiny speck of land)? What is not clear is whether he thought species actually were altered or whether he was only considering the possibility. It is also not clear whether by "new creation" he meant by transmutation or through special creation by the Creator, but probably the former. He was also saying that, if one species of rhea had in fact been altered to make two, the change had occurred abruptly (and presumably, therefore, suddenly) with no intermediates. This idea then became one of the foundations of his thinking, and it was a long time before he would abandon it.

Following this, Darwin wrote: "The same kind of relation that common ostrich bears to (Petisse . . .). Extinct guanaco to recent: in former case position, in latter time." The entry continues, "As in first cases distinct species inosculate, so must we believe ancient ones: (therefore) not gradual change or degeneration. From circumstances: if one species does change into another it must be per saltum. This representation of species important, each its own limit & represented . . . inosculation alone shows not gradation."[30]

This entry represents Darwin's first breakthrough. What he had done was a brilliant piece of associative reasoning, the kind that distinguished him in comparison with all his peers. He intuited that there was a similar relationship between species replacing each other in geographical space (that is, the northern and southern rheas) and replacing each other in time (the fossil and recent sloths and "camels"). It is interesting that he used MacLeay's term "representative." In MacLeay's system, "representative" species were those that occupied equivalent niches, had similar appearance and habits but had nonoverlapping distributions ("represent" in this case meant literally "take the place of").[31] The fossil and recent mammals Darwin had seen similarly represented each other in time. Darwin had also adopted the quinarian concept of "inosculation" (in which "osculant" species must be discretely different and arise *per saltum* [in leaps])—turning MacLeay's basic creationist system against itself.

Darwin inserted a reference to this new insight into the manuscript of *Journal and Remarks*. He termed this phenomenon of replacement of species in space and time the "law of succession of types [that] although subject to some remarkable exceptions, must possess the highest interest to every philosophical naturalist, and was first clearly observed in regard to Australia, where fossil remains of a large and extinct species of Kangaroo were discovered buried in a cave. In America the most marked change among the mammalia has been the loss of several species of Mastodon, of an elephant, and of the horse. These Pachydermata appear formerly to have had a range over the world, like that which deer and antelopes now hold. If Buffon had known of these gigantic armadilloes, llamas, great rodents, and lost pachydermata, he would have said with a greater semblance of truth, that the creative force in America had lost its vigour, rather than that it had never possessed such powers."[32]

At this stage Darwin had still been making largely *conditional* entries in his notebooks, on the basis of *"if"* one species does change into another. On page 153 of the Red Notebook, he became more positive. Returning to the subject of the rheas and the problem of gradual versus episodic change, he wrote: "When we see . . . two species. Certainly different. Not insensible change.—Yet one is urged to look to common parent?"

The next stage would involve a discrete shift in language, from "if species changed" to "how" and "why."

SIXTEEN

Notebook B

Darwin finished his manuscript of the *Journal and Remarks* around the end of July 1837. At almost the same time (sometime between the end of June and the beginning of August), he came to the end of Red Notebook and opened Notebook B, on (as he later wrote), " 'transmutation of Species.'—Had been greatly struck from about month of previous March—on character of S. American fossils—& species of Galapagos Archipelago.—These facts origin (especially latter) of all my views."[1]

Having decided that transformism (referred to henceforth as *transmutation*) was a real phenomenon, Darwin knew that he needed both to find the causal mechanism and to unravel the consequences in terms of a new biological worldview. Significantly, he began not with a philosophical analysis of abstractions like those of Lamarck, but with a long look at the causative aspects of Erasmus Darwin's ideas in *Zoonomia*. The main point that Darwin took from his grandfather's work, at least as recorded in notebook pages 1 to 14, was that any possible mechanism for transmutation had to be based in "generation." He followed Erasmus Darwin in distinguishing vegetative propagation (by cuttings and buds that simply continue the original stock) from reproductive generation which, through breeding of males and females, introduces the capacity for variation. New species would arise if, for example, populations were accidentally transported to an island, or otherwise isolated. "Let a pair be introduced & increase slowly, from many enemies, so as often to intermarry who will dare say what result. According to this view animals, on separate islands, ought to become different if kept long enough.—apart, with slightly different circumstances.—Now Galapagos Tortoises, Mocking Birds; Falkland Fox—Chilloe Fox."[2]

After he had made these entries, a number of explicitly transmutationist entries followed, discussing how such new species might be "propagated."[3]

> Propagation of species we can see why a form [apparently he meant "generic type"[4]] peculiar to continents; all bred in from one parent. (Notebook B, 12)
>
> Then (remembering Lyell's arguments of transportal) island near continent might have some species same as nearest land, which were late arrivals others old ones, (of which noneof the same kind had in interval arrived) might have grown altered. Hence type would be of the continent though species all different. In cases as Galapagos & Juan Fernandez. (10)
>
> Propagation explains why modern animals same type as extinct which is law almost proved.—We can see why structure is common to certain countries when we can hardly believe necessary, but if it was necessary to one forefather, the result, would be as it is. Hence Antelopes at C. of Good Hope. Marsupials at Australia. [Added later: "Will this apply to whole organic kingdom, when our planet first cooled."] Countries longest separated greatest differences—if separated from immense ages possibly two distinct type, but each having its own representatives—as in Australia. This presupposes time when no Mammalia existed; Australian Mamm were produced from propagation from different set, as the rest of the world. (14–15)
>
> This view supposes that in course of ages. & therefore changes. Every animal has tendency to change.—This difficult

to prove cats &c from Egypt no answer because time short &
no great change happened. (16)

The early pages of Notebook B show a man in a great hurry, carried
along by a rush of new ideas. By page 26 he had sketched out what an
evolutionary scheme would look like—a coral. It was the "coral of life,
base of branches dead; so that passages cannot be seen.—this again offers contradiction to constant succession of germs in progress.—no
only makes it excessively complicated." (By "passages," he meant the
continued existence of taxa intermediate between species or groups of
species.)

Darwin tested that thought against the quinarian view: "We may
fancy, according to shortness of life in of species that in perfection, the
bottom of branches deaden.—so that in Mammalia birds it would only
appear like circles:—& insects almost articulate.—but in lower classes
perhaps a more linear arrangement" (Notebook B, 27).

But he was still worrying about gaps in diversity and the apparently short lifespan of highly evolved species. "There does appear
some connection shortness of existence, in perfection species from
many therefore changes and base of branches being dead from which
they bifurcated" (Notebook B, 29). Six pages later, still worrying away
at finding a lawful system of nature, he asked, "Is this shortness of
life of species in certain orders connected with gaps in the *series of
connection?*"

On page 36 he recorded yet another major breakthrough—revising
his coral analogy into his famous branching tree diagram, showing the
relationships among a hypothetical group of four living and twenty-one
extinct taxa. Although Lamarck had given a rudimentary sense of a
branching relationship among major groups of animals in his *Philosophie zoologique,*[5] no one had previously seen relationships as treelike
and therefore explicitly genealogical. Every subsequent depiction of
evolutionary relationships among organisms in the Darwinian era can
be seen as a version of this, already quite complex, diagram.

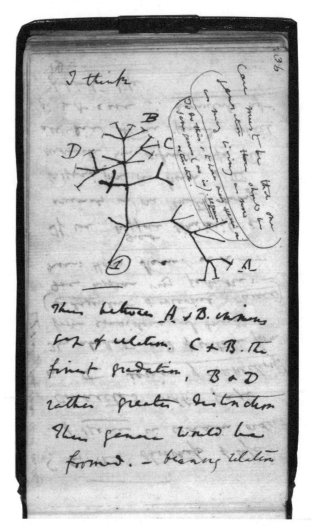

3. Page 36 of Darwin's Notebook B, where he first sketched his "tree of life" metaphor for divergent evolution. In this scheme, the unlabelled twigs are extinct, leaving living species more (A, D) and less (B, C) distant from each other. (Courtesy of Cambridge University Library [MSS.DAR 121:36].)

The essence of Darwin's tree diagram was that, as species evolved one from another in a branching pattern, some would become extinct, leaving gaps that would account for the clustering of some species together (a genus), separated from others.

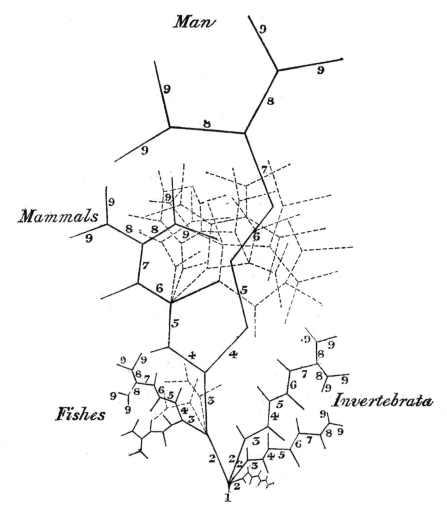

4. Martin Barry's "Tree of Animal Development," showing an idealized pattern
of commonality and difference in developmental history. In this scheme, low
numbers indicate the (early) acquisition of major characters such as class (2) and
order (3), which precede the (late) acquisition of species level (6), sexual
differentiation (8), and individual (9) differences. *Edinburgh New Philosophical
Journal* 22 (1837): 345. (Courtesy of the Ewell Sale Stewart Library, Academy of
Natural Sciences of Philadelphia.)

Remarkably, Darwin's tree bears a resemblance to a diagram in a work by Dr. Martin Barry published at just the same time in the familiar *Edinburgh New Philosophical Journal*.[6] Barry was an expert on reproduction and early embryological development. In a long paper entitled "On the Unity of Structure in the Animal Kingdom," he argued in large part from the observations of the German anatomist and embryologist Karl von Baer, who (following yet another German embryologist, Johann Friedrich Meckel), had built a quasi-evolutionary system on the basis of the observation that the embryological stages of different animals bore a greater relationship to each other than did the adults: the earlier the stage, the greater the similarity. Von Baer tried to extend the argument to suggest that the embryological stages in any one animal's development represented the succession of the adult stages of its ancestors. This eventually became the familiar (and erroneous) theory of recapitulation summed up in the phrase "ontogeny recapitulates phylogeny." However interpreted, these similarities pointed to the fact that differences in structure among animal groups were acquired sequentially, and that the similarities were genetic, representing a continuity of genealogy.

The question of what greater meaning lay behind the work of Meckel and von Baer (and also Etienne Serres) was a major topic of discussion in the late 1830s. One of those particularly impressed by the German ideas was Darwin's new friend Richard Owen, who at first dismissed it as "nonsense" (Notebook B, 163). Later Owen came to see the power of the idea.[7] Darwin noted: "Owen says relation of Osteology of birds to Reptiles shown in osteology of young Ostrich."[8] Darwin was intrigued later by Goethe's insight that the vertebrate skull consisted of fused, modified vertebrae: "The head being six metamorphosed vertebrae, the parent of all vertebrate animals.—must have been like some molluscous animal with a vertebra only & no head—!!"[9]

Barry's interest in this was not overtly evolutionary. Like Cuvier and Geoffroy, and many philosophical zoologists of the time, he was concerned with finding the commonality of structure (and function) of different animal groups. In formal morphology, there is a vertebrate plan; subgroups like birds show modified versions of the ground plan; woodpeckers and swans show their own derived versions of the bird plans, and so on. "Unity of type" therefore stood at the core of "affinity" and, therefore, classification.

Geoffroy was also someone who influenced Darwin, who made several references to his *Principes de philosophie zoologique.* Geoffroy argued that homologous organs arose only when genealogically related animals had descended from a common ancestor. The wings of bats and birds, for example, were only analogous as wings, but homologous at the level of (derived) vertebrate forelimbs.[10]

In reviewing the phenomenon of resemblance among fetal stages, Barry concluded that, in the development of each individual, "a heterogeneous or special structure, arises only from out of more homogenous or general" and that structures arise in individual development "in the order of their generality." Basically, "all the varieties of structure in the animal kingdom, are but modifications of, essentially, one and the same fundamental form."[11] Therefore, in the development of a bird, common animal features appear in the earliest embryological stages, followed by features characteristic of vertebrates, then of tetrapods, then of reptiles, then of the particular kind of bird.

Barry's drawing showing relationships among mammals, fish, and invertebrates (figure 4) looks similar to Darwin's branching diagram, but the two are quite different in concept. Barry's was intended to show the divergent pathways in the development of the characteristics that define, in turn, individuals, sexes, varieties, species, genera, family, orders, and classes. It was presented as a set of formal relationships dictated by knowledge of structure and classification. Darwin's diagram, by contrast, shows relationships among species, rather than a taxonomic hierarchy. If anything, Darwin's diagram represents the end branches in Barry's scheme. Darwin does not refer in the notebooks to Barry's paper, but he almost surely saw it, and it seems likely that Barry's diagram at least gave him a model for his own very different scheme.[12]

Darwin's first note about the body of work of by Serres, Meckel, and von Baer comes 127 pages after his branching diagram. There is much debate in the Darwinian literature about Darwin's debts to the various versions of this body of work, but there can be no doubt that its demonstration of some kind of parallel between individual development in animals and phylogeny impressed him strongly. "Originality is given (& power of adaptation) is given by *true* generation, through the means of every step of progressive increase being initiated in the womb, which has been passed through to form, that species" (Notebook B, 78).

Thus far, Darwin had committed himself to an evolutionary philosophy for his "new system of nature." He adopted his grandfather's (and Buffon's) sense of the central importance of "generation" (we might say genetics and development) both as a cause of stability within species and as a potential source of change. Critically, he had seen that geography, particularly when under change through geology, was a major factor in the processes of change. And he had articulated the key proposition that transmutation of species proceeded in a branching fashion that resulted in the emergence of divergent species. But he was a long way from having a comprehensive theory. The early entries in Notebook B show that, while Darwin was already fully committed to discovering a new law of transmutation, he had a number of tricky issues to reconcile on the way. There were prior conceptions that had to be examined and either incorporated or discarded. It is impossible, therefore, to trace a single, direct line in the evolution of Darwin's ideas.

As his primary data were in the classification and geographical distribution of species, Darwin remained engrossed in ways of explaining and reinterpreting MacLeay's quinarian system, which he obviously still thought had some kernel of merit. (There is no evidence that Darwin ever found the number five per se persuasive; some scholars had already revisited MacLeay to found rival systems upon the numbers four or three.)

For a while, Darwin thought that there might be a more dynamic basis for MacLeay's patterns of affinity and analogy. He toyed with the notion that each group would show a "triple branching . . . owing to three elements air, land & water, & the endeavour of each one typical class to extend his domain into the other domains. & subdivision three or more . . . if each Main stem of the tree is adapted for these three elements, there will be certainly points of affinity in each branch"

(Notebook B, 23–24). In any group there would be members adapted for life on land, air, water, with one of the three predominating. Thus, among mammals there are predominantly land forms, with some aquatic and some aerial. The porpoises have shapes analogous to those of sharks; bats have wings analogous with those of birds.[13]

Twenty pages later, he started an entry on Quinarism with what was, by then, a surprising reference to God. "The Creator has made tribes of animals adapted preeminently for each element, but it seems law that such tribes, as far as possible, compatible with such structure are in minor degrees adapted for. other elements. every part would probably be not complete, if birds were fitted solely for air & fishes for water. If my idea of origin of Quinarian system is true, it will not occur in plants which are in far larger proportion terrestrial" (Notebook B, 44, 45).

Darwin's views continued to be colored by MacLeay's view that "osculant" species were necessarily distinct from other species, which therefore required that, if there had been any transmutationist change, it had to be per saltum rather than by gradation. His entry on isolation (on page 7 of the Red Notebook, quoted above) continued: "As we thus believe species vary [in] changing climate we ought to find representative species; this we do in South America closely approaching.—but as they inosculate, we must suppose the change is effected at once." (Again, the word "representative" was used in MacLeay's sense of "taking the place of.") And on following pages he compared this with the view that "Species according to Lamarck disappear as collection made perfect" (Notebook B, 7–12). It will be noted that there is a potential conflict here between the view that species vary and may change in isolation "if kept long enough" and the assumption that the origin of a new species occurs suddenly, per saltum. It took a long while for Darwin to sort out this inconsistency.

Another of Lamarck's key concepts—monadism—pops up regularly in Notebook B.[14] Lamarck thought that new lineages of organisms arose through spontaneous generation of primitive monads. At the beginning there were separate monads for the animal and plant kingdoms, but Lamarck also allowed the possibility of new monads having arisen "at the beginning of certain branches of those scales."[15]

In any evolutionary scheme, there is the problem of explaining why at present we have both apparently primitive and advanced taxa alive. In Lamarck's scheme, this was explained if monads were not produced all at the same (ancient) time, but new ones were steadily called forth to evolve, in turn, in progressive fashion toward greater complexity (higher up the Chain of Being). For Lamarck, the different state of advancement of, say, modern worms and mammals was due to their monads having been created (by spontaneous generation) at different times.

There is much that is confusing in Lamarck's ideas. For him, the key differences among groups arose in these lineages (evolution of legs, lungs, hair, and so on in vertebrates). But he did not specifically identify such major phylogenetic changes with the transmutation of species produced by local adaptation to environmental conditions or through use and disuse. He did not project transmutation as a *constantly* branching process or as the driving agent of phylogenetic change.

Darwin seems, for a while, to have been somewhat trapped by the monad theory. The concept of monadism showed up principally in the form of the idea that the monad (by which he seems also to have meant the *lineage* arising from a first germ) had a finite existence. Thus, he wrote: "If we suppose monad definite existence, as we may suppose is the case, their creation being dependent on definite laws, then those which have changed most, owing to accident of positions must in each state of existence have shortest life. Hence shortness of life of Mammalia" (Notebook B, 22–23).[16] However, by page 29, he had decided that "monad has not definite existence" because all the animals "springing from the same branch" would all die at once, "which is not the case" (35).

This shows that Darwin was trying to imagine the consequences of transmutation across a wide spectrum of subject, from patterns of diversification to homology and analogy, numbers of species, numbers of species per higher group, longevity of groups, and patterns of extinction—all of which combined to affect biogeographical patterns.

At this stage, Darwin believed conventionally that all living forms were perfectly adapted to their particular circumstances (a leftover of Paleyian natural theology) although, on pages 38 and 39 of Notebook B, he had also posited that extinction came through lack of adaptation. This was a problematic issue because of his belief that the large mammals of North and South America had not been driven to extinction by environmental change or other nonadaptedness. However, by page 64, Darwin had decided that individuals die in order both "to perpetuate certain peculiarities, (therefore adaptation), & to obliterate accidental varieties, & to accommodate itself to change. . . . Now this argument applies to species.—If individual cannot procreate, he has no issue, so with species."

As his ideas started to come together, he was still using phrases like "creative force" but now in quite a generic sense. He did not mean specifically "the Creator." For example, he wrote: "How does it come wandering birds, such as sandpipers not new at Galapagos.—did the creative force know that these species could, arrive—did it only create those kinds not so likely to wander" (Notebook B, 100).

Confident of his view that new species arose when geographically isolated and by adapting to new conditions, Darwin dismissed the idea that God ordered every change among organisms: "Much more simple, & sublime power let attraction act according to certain laws such are inevitable consequence let animal be created, then by the fixed laws of generation, such will be their successors.—let the powers of transportal be such & so will be the form of one country to another.—let geological changes go at such a rate, so will be the numbers & distribution of the species!!" (Notebook B, 102). And he added further elements to his argument: "It may be argued representative species chiefly found where barriers & what are barriers but interruption of communication, or when country changes" (103). New species arising from "fresh creations" was "mere assumption, it explains nothing further, points gained if any facts are connected" (104).

Later on, he rehearsed this argument, recognizing the difficulty: "Before attraction of Gravity discovered, it might have been said it was as great a difficulty to account for movement of all, by one law. As to account for each separate one, so to say that all Mammalia, were born from one stock, & since distributed by such means as we can recognize, may be thought to explain nothing.—it being as easy to produce for the Creator two quadrupeds at South America Jaguar & Tiger & Europe to produce the same one" (Notebook B, 196). A notable feature of this entry is his assumption that his theory would eventually stand on equal terms with Newton's theory of gravitation.

Soon the change away from the Creator had been made complete: "Has the Creator since the Cambrian formations gone on creating animals with same general structure.—miserable limited view" (Notebook B, 216). Darwin's growing confidence in his ideas was reflected in his language. He dismissed Lamarck's idea that change in part occurs through the animal willing it as "absurd" and used the same adjective to describe MacLeay and every version of the Quinary System.

Finally, on page 219 of Notebook B, he started at last to refer to "my theory." It would be the kernel of a new system of nature and "give zest to recent & Fossil Comparative Anatomy. & mind heredity, whole metaphysics" (228). No doubt remembering his friend Browne at Edinburgh, he was still conservative enough to note, however, that "the soul by consent of all is superadded, animals not got it . . . [but] if we let conjecture run wild then animals our fellow brethren in pain, disease, death & suffering . . . they may partake, from our origin in one common ancestor we may all be netted together" (232).

Darwin's new system consisted of both a theory of transmutation and a view of its greater significance (extinction, distribution/centers of creation). He still had not concluded that the origin of species must occur through gradual change, even though everything in Lyell's "uniformitarian" view of the earth suggested it. He seemed to be edging toward that view, however. On page 154, using the analogy with artificial selection, he had dismissed the notion that one ought to find all the intermediates between species: "When humans breed new races, they discard the intermediates." One of the later entries in this notebook noted that the reason that human breeding of plants and animals (artificial selection) had not produced new species is that there had not been enough time (244). He argued by the same analogy that "man has not had time to form good species" out of the various human races. The modern origin of the human species was demonstrated by there being only one species.

Darwin made his final entry in Notebook B around February or March 1838. In some eight hectic months, essentially as a side issue while also engaged in his various publishing projects, Darwin had made enormous strides toward working out a theory of transmutation. Both in terms of a causal mechanism and the logical consequences of that mechanism for explaining the data of biological and paleontological diversity, there would still be a long way to go before he would be satisfied. Even more work had to be done before he would feel confident in sharing his ideas with others.

Meanwhile, his double life continued.

SEVENTEEN

Moving Forward, Living a Lie

After a year in London, Darwin's life had become unbearably complicated. While the social side of life was stimulating, both in hobnob-
bing with great scientists and enjoying the company of literary types
at the salon that Erasmus maintained, it was physically and mentally
exhausting. The worst part of it was that, while enjoying the company
of the capital's scientific men, in secret Darwin was living his lie,
never able to admit to investigating the subject that his heroes and
his friends all considered to be total nonsense at best and heresy at
worst.

At this stage of his theorizing, he had nothing solid enough to ex-
pose to the critical scrutiny of people like Lyell and Sedgwick—
known for their vehement opposition to any kind of transmutationist
views. It was enough that Darwin was making waves by challenging
geological orthodoxies. So Darwin was forced to carry on a dual existence. In public he was the promising young geologist, a gifted naturalist and writer, and a mainstream scholar whose ideas were well
supported by Lyell and others. In private he became a subversive,
feverishly jotting down thoughts for which he knew everyone would
condemn him. If those thoughts were to get out before he was ready,
his new position in the world of English natural historians would be
irrevocably damaged. Not surprisingly, therefore, his health started to
break down once again.

The problems started early on. In September 1837, he frankly admitted to friends, including Fox and Henslow, that "I have not been
well of late with an uncomfortable palpitation of the heart."[1] When
pressed to accept the offer of the secretaryship of the Geological Society, he begged off because: "I doubt how far my health will stand, the
confinement of which I have to do without any additional work . . .
when I consulted Dr Clark in town, he at first urged me to give up entirely all writing and even correcting press for some weeks. Of late, anything which flurries me completely knocks me up afterwards and brings
on a bad palpitation of the heart."[2] Perhaps the key phrase in this last is
"anything which flurries me": Darwin's old nemesis.

Darwin's health problems were not confined to irregularity of the
pulse. While the skin eruptions seem to have eased in frequency, his
stomach began acting up. To his sister Caroline, now married to her
cousin Josiah Wedgwood, he wrote: "I find the noodle & the stomach
are antagonist powers, and that it is a great deal more easy to think too
much in a day, than to think too little—What thought has to do with di-
gesting roast beef,—I cannot say, but they are brother faculties. I am living quietly, and have given up all society."[3] These new symptoms were
largely abdominal in presentation. They were quite different from, but
additional to, his eruptions of "eczema" and palpitations.

In June, hoping to improve his health, he made a trip to Scotland to
"take a solitary walk on Salisbury crags . . . to geologise the parallel
roads of Glen Roy,—thence to Shrewsbury, Maer for one day, & London for smoke, ill health & hard work."[4] His doctors had prescribed a
time in the country; Darwin, typically, would spend it working as much
as relaxing. It seems most odd, however, that Darwin, who had evidently
suffered so dreadfully from seasickness on the *Beagle,* should have em-
barked for Edinburgh on a steamer. The North Sea route from London
to Edinburgh by sea traverses some of the roughest water one could find
without returning to Cape Horn. No doubt Darwin was no great fan of
the stagecoach, either. As it happened, on this occasion he flourished
(and boasted): "My trip in the steam packet was absolutely pleasant, &
I enjoyed the spectacle, wretch that I am, of two ladies & some small

children quite sea sick, I being well. Moreover on my return from Glasgow to Liverpool, I triumphed in a similar manner over some fully grown men."[5]

While this extended trip seemed to have had quite straightforward aims, it actually had several agendas. It is interesting, for example, that, twelve years after having left Edinburgh and the influence of Robert Jameson, he felt the need to take his newfound skills in geology and to settle for himself the question of the trap dykes at Salisbury Crags. He wrote to Lyell: "I want to hear, some day, what you think about that classical ground:—the structure was to me new & rather curious,—that is if I understand it right."[6] This statement rather contradicts his recollection in the later *Autobiography* of his visit there as a student with Jameson when the structure was supposedly perfectly clear.[7] In his Glen Roy Notebook, he made a sketch of the famous intruded dykes with the annotation "Veins, amygdaloidal," a note that presages the scornful "amygdaloid" of the *Autobiography*.[8] A separate set of notes on Salisbury Crags, preserved at Cambridge University Library, seems to indicate that he planned a publication on the subject, but he evidently shelved it.

— The parallel roads of Glen Roy presented something of a greater geological puzzle. Like Salisbury Crags, the phenomenon seemed ideal for the young man to show his geological mettle. The sides of valley of the River Roy in the western Highlands above Fort William, are carved into terraced shapes—the parallel roads. Darwin proposed to explain these former marine beaches, raised up serially in a set of Lyellian elevations of the land. "Here I enjoyed five days of the most beautiful weather, with gorgeous sunsets, & all nature looking as happy, as I felt.—I wandered over the mountains in all directions & examined that most extraordinary district.—I think without any exception,—not even the first volcanic island, the first elevated beach, or the passage of the Cordillera, was so interesting to me, as this week. It is far the most remarkable area I ever examined.—I have fully convinced myself, (after some doubting at first) that the shelves are sea-beaches,—although I could not find a trace of a shell, & I think I can explain away most, if not all, the difficulties."[9] Writing his subsequent paper on the subject was one of Darwin's "most difficult & instructive tasks," and it was to prove a notable failure.[10]

There were agendas beyond science for Darwin's Scottish trip of 1838. The young Darwin was considered a fine potential catch in the swirling matrimonial waters of Cambridge and London society. Darwin himself had been wondering about the advisability of marriage, and the wife of the geologist Leonard Horner may have precipitated things a little. Charles Lyell was married to one of the five Horner daughters and fancied Darwin for a brother-in-law (later Lyell's brother Henry married one of the daughters). It does not take much imagination to feel the pressures on Darwin. If he married into the Horner family, he would be even more well connected to the scientific world of London and beyond. Having had little contact with marriageable young women for six years, Darwin flirted and dithered, alternately bold and retiring.

Darwin used the opportunity of a week's stay in Shrewsbury on the way back from Scotland to consult with his father on the subject of matrimony, and he jotted down some notes on the back of a letter.[11] If he did not marry, he thought, there would be opportunity for travel: "exclusively geological United States, Mexico," unless he found himself becoming more of a zoologist. If he did not travel, he would have time for "work at transmission of species." If he married, money would be short. Perhaps he could get a professorship at Cambridge; if not, he would be a "poor man; outskirts of London, some small Square." But he really wanted to live in the country. On balance, these notes argued against marriage.

The final agenda item for the Scottish trip was that, while in Shrewsbury, Darwin would make a trip over to Maer, home not only of his favorite uncle but his unmarried female cousins. The shade of Jane Austen's Mrs. Bennett was indeed hovering over the house at Maer as well as Mrs. Horner's.

Some three months later, he revisited the subject of matrimony and laid out his now-famous balance sheet, "This is the Question," as two columns of notes, headed "Marry" and "Not Marry." Among the arguments he mustered were: "Constant companion, (& friend in old age) who will feel interested in one." Against this could be measured: "Not forced to visit relatives, & to bend in every trifle.—to have the expense and anxiety of children—perhaps quarrelling." There were the "charms of music & female chit-chat" versus the emphatic "*Loss of time.*" "Perhaps my wife wont like London; then the sentence is banishment degradation into indolent, idle fool" versus "Better than a dog." It is hard to read this without thinking that Darwin must either have had a much better sense of humor than he had hitherto revealed or that he was an utter boor. On the whole, one must perhaps give credit to his wit.[12] On balance, Darwin pictured himself with "a nice soft wife with a sofa and a good fire, & books & music perhaps" and concluded "Marry—Marry—Marry Q.E.D."

Even so, he had many qualms: "an infinity of trouble & expense in getting & furnishing a house . . . morning calls . . . how should I manage all my business if I were obliged to go every day walking with my wife . . . never mind my boy—Cheer up—One cannot live this solitary life, with groggy old age, friendless & cold, & childless staring in ones face, already beginning to wrinkle.—Never mind, trust to chance—keep a sharp look out—There is many a happy slave—."

At some point, he determined to propose to Emma, the youngest of the Wedgwood children. A year older than Darwin, at age thirty, she probably had few hopes left of marriage. Darwin made the decision for himself, but it was the match that both families had secretly been hoping for.

On the scientific front, his work on transmutation had accelerated rapidly. Nonetheless, if his personal journal is to be trusted, he still worked "on Species" only intermittently—in February and May 1838, February 1839, December 1839 to February 1840, and so on.

In Notebook C, Darwin's entries were focused on transmutation and the subject that, for Darwin, was closely related—the long-range dispersal of animals and plants (and consequently the barriers that prevented dispersal). By every indication, he was closing in on his "idea," but caution must be urged. Although Darwin now began many entries in the notebooks with the words "In my theory" or "If my theory right," he still did not state precisely what his theory was. While we know what the later versions of his theory was, we may be quite sure that the early and intermediate versions were different; we just don't know exactly what they were. It would be a mistake to assume that Darwin had a clear blueprint right from the start.

From the evidence of the notebooks, one can see that Darwin continued to proceed, intellectually, on a very broad front. Notebook C shows him for the first time considering artificial selection as an analogue of what occurs in nature. But on page 177, he wrote that his theory had "broadly scarcely any novelty . . . only slight differences, opinion of many people in conversation, the whole object of the Work is its proof, its limiting, the allowing at same time true species & its adaptation to classification & affinities, its extension."

Notebook C is remarkable for the range of scientific works referred to. Darwin was reading even more prolifically than ever and commenting on a huge range of subjects, from basic exploration and travel to the mind and the inheritance of instinct; from the relationship between species diversity in a group and geography and phylogeny to hybrids and animal and plant breeding.[13] There was also a strong concentration of speculation about the significance of the races of humans, due in part to his reading of James Cowles Prichard's *Researches into the Physical History of Man* (third edition, 1837), in which the author speculated that the cause of the differences among the races was adaptation to different environmental conditions. Prichard did not believe that skin color was the sort of use-disuse character that Lamarck envisaged. No matter how deep a tan one acquired, one did not pass it on to one's children. Instead, what was inherited was a genetic predisposition to a particular level of tanning. Prichard saw the races of humans as an experiment in selective breeding. There is an interesting connection here to Darwin's Edinburgh years, when the first edition of Prichard's book (1813) was influential on the anatomist and physiologist William Lawrence, whose work Darwin read then.[14]

In a word, judging from the breadth of reading and the references that Darwin recorded in the notebooks, he was still *searching*—searching for the full matrix and outer boundaries of a theory rather than just polishing the version that we know.

"My theory" was, moreover, not just about transmutation and explanations of biogeographical patterns. One of his preoccupations all along had been with the evolution of mind, instinct, and reason. The subject also harked back to his student days at Edinburgh and the debate over whether the soul was a material phenomenon, contingent upon the structure of the brain, or something extramaterial. This led him to the works of David Hume and a second Scottish philosopher, Dugald Stewart. In July of 1838 he opened Notebook M, dealing with "metaphysical issues and expression," which was followed in October by Notebook N on the same subject.

He also continued to worry away at MacLeay's Quinarism, largely because, in explaining adaptation, he needed to sort out the significance of affinity and analogy. The quinary approach forced the researcher to look for causes in the patterns of taxonomic arrangement. For example, on page 61 of Notebook C, Darwin stated that "analogy may chiefly be looked for in the aberrant groups" and gives the instance of a walking flycatcher. He then segued neatly from taxonomy to evolution and sexual selection by following this sentence with a question: "Whether species may not be made by a little more vigour being given to the chance offspring who have any slight peculiarity of structure hence seals take victorious seals . . . hence males armed & pugnacious."

By page 170, he was bemoaning the "absurdity of Quinary arrangement," and he singled out Swainson's versions of it for particular contempt.[15] But he continued to use the language of Quinarism. For example, as late as Notebook C, page 207, he postulated that "aberrant forms produced where many species [*osculant* deleted]."

Another of his necessary preoccupations was heredity, of which none of the facts that are now familiar to every grade schooler were then known. The subject was central to the subjects of variation and adaptation, and also to the refutation of Lamarck's theory of inheritance of acquired characteristics. Here the field of animal and plant breeding was crucial to him, as it provided empirical demonstration of the heritability of variant characters. On page 66, Darwin noted, "With respect to my theory of generation, fact of armless parent not having armless child, shows there is reference to more than offspring (like atavism) & shows my view of generation right?" (But the reader will note the final question mark.)

As to how new adaptations arose, Darwin still had a quasi-Lamarckian view of the origin of novelties: "All structures either direct effect of habit, or hereditary & combined effect of habit—perhaps in process of change." He noted, with emphasis, "WHOLE race of that species must take to that particular habit." But he resoundingly rejected Lamarck's idea of the role of volition in producing change: "Lamarck's willing absurd."[16]

All this harks back to an old question, one that David Hume had discussed in *Dialogues concerning Natural Religion,* for example: do new structures—webbed feet, for example—arise before their use, or does use precede the structure? In the confusing language of the preceding quotation, here Darwin seems to be heading in the wrong direction. Selection (use) cannot act until after the variant condition has arisen to be used (and further reinforced by selection). In principle, contra Lamarck, no amount of use would *produce* a variant structure. Three pages later, he gave a nice example of his views, in which Lamarckian use and disuse is absent: "If puppy born with thick coat monstrosity, if brought into cold country and there acquired then adaptation" (by "acquired," he means propagated in a population).

Darwin continued to make notes about humans. On April 1, he visited the London Zoo and saw the popular orangutan display, in which these apes were treated as children. He wrote to his sister Susan, "I saw also the ouran-outang in great perfection: the keeper showed her an apple, but would not give it her, whereupon she threw herself on her back & after two or three fits of passion, the keeper said, 'Jenny, if you will stop bawling & be a good girl, I will give you the apple.'—She certainly understood every word of this & though like a child, she had great work to stop whining, she at last succeeded, & then got the apple, with which she jumped into an arm chair & began eating it with the most contended countenance imaginable."[17] In his notebook, he repeats these comments and, in this private world, dared to add a direct comparison with uncivilized human races: "Let man visit Ourangoutang in domestication, hear expressive whine, see its intelligence when spoken to . . . let him look at savage roasting his parent, naked artless, not improving yet improvable & let him dare to boast of his proud preeminence.—not understanding language of Fuegian, puts on par with Monkeys."[18]

Darwin was well aware of the problems his theory did not answer. One difficulty was to explain why differences among varieties were not swamped out by interbreeding. One of Darwin's first references to human races came in Notebook B, when he observed that the distinctness of the various tribes in Tierra del Fuego was due to reluctance to "cross readily."[19] Early in Notebook C, Darwin noted: "The most hypoth: part of my theory that two varieties, of many ages standing, will not breed together."[20] On page 201 he was resigned that "I fear argument must rest upon analogy & absence of varieties in a wild state."

In another entry, he admitted the difficulty of explaining the origin of major organs, like the eye, by accumulation of minor changes. "We may never be able to trace the steps by which the organization of the eye, passed from simpler stage to more perfect, preserving its relations.—the wonderful power of adaptation given to organization.—This really perhaps greatest difficulty to whole theory."[21] In an unconsciously humorous demonstration of his catholic reading, the following sentence reads: "There is breed of tailless cats near Bath."

Adding to Darwin's psychological stress and workload, in September, just when he was agonizing over whether to propose marriage to Emma, one of the last pieces of the transmutationist jigsaw puzzle fell into place.

Darwin already had a view of species changing under situations like isolation and under changed environmental conditions. He had seen that the history of life could be depicted as a set of branching trees. From his grandfather he had seen that sexual reproduction (generation) was the key to the development and maintenance of varieties.

He knew of the power of artificial selection (animal and plant breeding) and saw its value as an analogy for what happens in nature. He had
not catalogued the extent and range of natural variation in populations
of species (the raw material upon which his principle of natural selection would act). He did not know what caused variation, and he did not
know the driving cause of change.

It was time for a new notebook, and at the front of Notebook D,
Darwin wrote: "Towards close I first thought of selection owing to
struggle." According to his *Autobiography,* the facts of island biogeography, such as the species diversity on the Galapagos, and the general
phenomenon of adaptation had long "haunted" him. But he could find
no vera causa element on which to base a theory of the cause of change.
He lacked the "smoking gun" of modern metaphor, and "it seemed
to me almost useless to endeavour to prove by indirect evidence that
species have been modified."[22]

He jotted down some ideas for a "Theory of Geograph. Distrib: of
organic beings." As a start, he postulated that "animals of same classes
differ in different countries in exact proportion to the time they have
been separated; together with physical differences of country."[23] He
noted that "line of Rocky mountains separate almost all mammals of N.
America & many birds."[24] In the area of "domesticated productions,"
Darwin's reading continued to be eclectic, and his imagination produced a whole range of new ideas, often even before he had properly
fixed a basis for them.

The notebook entries continued to vary from trivial to fundamental:
"What a magnificent view one can take of the world Astronomical causes,
modified by unknown ones. Cause changes in geography & changes of
climate superadded to change of climate from physical causes.—these su-
perinduce changes of form in the organic world, as adaptation. & these
changing affect each other, & their bodies by certain laws of harmony
keep perfect in these themselves. . . . How far grander than idea from
cramped imagination that God created. (warring against those very laws
he established in all organic nature)."[25]

"Is there some law in nature an animal may acquire organs, but lose
them with more difficulty."[26] "It is important with respect to extinction

of species, the capability of only small amount of change at any one time."[27] And he was still very modest: compared with Humboldt, Geoffroy de Saint Hilaire, and Lamarck, "I present no originality of idea— (though I arrived at them quite independently & have used them since) the line of proof & reducing fact to law only merit if merit there be in following work."[28]

His notebook entries continued to be heavily concerned with "generation" but, as there was no theory, or even rudimentary science of genetics, to support his theory of nature, Darwin had to rely upon example after example from plant and animal breeding, and principles of sexual reproduction, trying to find some commonalities. A key issue was "what characters chance to be hereditary whether important or not."[29]

Then, in "October 1838," he picked up a copy of Thomas Robert Malthus's book on population, with its mathematical demonstration that a struggle for survival inevitably resulted from the capacity of every species to overproduce offspring. "Being well prepared to appreciate the struggle for existence which everywhere goes on from long-continued observation of the habits of animals and plants, it at once struck me that under these circumstances *favourable variations would tend to be preserved, and unfavourable ones to be destroyed.* The result of this would be the formation of new species. Here, then, I had at last got a theory to work by" (emphasis added).[30]

From internal evidence, the notebook entries suggest that he had read Malthus no later than September, not October. He did not note the momentous event in his personal diary and the entry for September 14, 1838, even states that he had "frittered these foregoing days away in working on Transmutation theories." Perhaps it is also significant that, having read Malthus and also decided to ask Emma to marry him, he noted: "*Lost* 6, 7, 8th of Novemb. unwell."[31]

Malthus brought home to him the potential importance of superfecundity—the fact that individuals produce during their lifetimes many times more offspring than the two that are needed for stable population size. Most offspring never contributed to the next generation but were lost, principally as juveniles. A female cod might produce more than a million eggs in its lifetime. Out of the many individual variants that all species have the capacity to (over)produce, the necessities of number—the "shifting oeconomy of nature"— selected the most favored individuals—and races and species—for survival. What Malthus had shown was inevitable, for human populations did not just create an *analogue* of artificial selection—it established the mathematical certainty of a parallel *natural* selection.

Darwin had, of course, read the elements of Malthus's argument years earlier in Paley's *Natural Theology*.[32] On page 64 of Notebook B (more than a year previously), he had posited that individuals "die, to perpetuate certain peculiarities, (therefore adaptation), & to obliterate accidental varieties, & to accommodate itself to change." And, "With respect to extinction, we can easily see that a variety of the ostrich, Petise, may not be well adapted, and thus perish out; or on the other hand, like Orpheus, being favourable, many might be produced."[33]

The emphasis that Malthus gave to the force of numbers gave teeth to a "struggle for existence" and finally provided a mechanism.[34] Darwin's first reference to Malthus came near the end of Notebook D in the note: "the warring of species as inference from Malthus."[35] At once, he developed a new analogy to describe the processes of adaptation: "One may say there is a force like a hundred thousand wedges trying to force into every kind of adapted structure into the gaps in the oeconomy of Nature, or rather forming gaps by thrusting out weaker ones. The final cause of all this wedgings, must be to sort out proper structure & adapt it to change.—to do that, for form, which Malthus shows, is the final effect of this populousness."[36]

Darwin noted on the flyleaf of Notebook D "1838 . . . towards close I first thought of selection owing to struggle." However, while Malthus showed the mathematical inevitability of a struggle for existence, the struggle itself was a well-known phenomenon. It appeared, as we have seen, in Paley. The Swiss botanist de Candolle (a plant geographer very much in the style of Humboldt) in 1820 stated, "All the plants of a given country are at war with another . . . the more prolific gradually make themselves masters of the ground."[37] Darwin had mentioned this work as early as page 280 of Notebook B and may have come to de Candolle through Lyell, who had quoted him and developed his ideas in the second volume of his *Principles of Geology*. Similarly, Tennyson (*In Memoriam*) in 1850 famously described nature as "red in tooth and claw . . . from scarped cliff and quarried stone / She cries, 'A thousand types are gone: / I care for nothing, all shall go.' "

Darwin began Notebook E at the beginning of October 1838 and continued to make entries in it until July 10, 1839. He opened with a continuation of his thoughts on Malthus: "It accords with the most liberal spirit of philosophy to believe that no stone can fall, or plant rise, without the immediate agency of the deity. But we know from experience! That these operations of what we call nature, have been conducted almost! Invariably according to fixed laws: And since the world began, the causes of population & depopulation have been probably as constant as any of the laws of nature with which we are acquainted.—This applies to one species—I would apply it not only to population & depopulation, but exterminations & production of new forms—their number and correlations."[38] With the insights triggered by Malthus, Darwin's theory was approaching a coherent shape.

Throughout Notebook E, Darwin continued to mix apparently trivial facts and quotations with deeper insights, but the analytical entries began to prevail. Still thinking about Meckel, von Baer, Barry, and the way in which the embryonic history of an individual seems to parallel the phylogenetic history of its lineage, he again referred to the work of Owen: "relation of Osteology of birds to reptiles shown in osteology of young ostrich."[39] On page 51, he mused on the broader implication of his ideas: "Thinking of effects of my theory, laws will be discovered. Of co relation of parts, from the laws of variation of one part affecting another."

By now, Darwin had got a good handle on the subject of variation and the analogy between artificial and natural selection. "It is a beautiful part of my theory, that domesticated races of organics, are made by precisely same means as species—but latter far more perfectly & infinitely slower."[40]

And by now he was sure that all transmutation was gradual. Change per saltum had definitely been abandoned. "Varieties are made in two ways—local varieties, when whole mass of species subjected to some influence, & this would take place from changing country: but greyhound. And pouter Pidgeons race horse have not been thus produced, but by the training, & crossing & keeping breed pure. . . . Has nature any process analogous—if so she can produce great ends—But how. . . . Make the difficulty apparent by cross-questionning.— even if placed on Isld—if &c &c.—Then give my theory. Excellently true theory."[41]

At the same time, having read a collection of Hume's works, Darwin noted in his second Metaphysical Notebook (Notebook N) that: "Hume has section . . . on the Reason of animals . . . also on origin of religion or polytheism . . . however, he seems to allow it is an instinct. . . . I suspect the endless round of doubts & skepticisms might be solved by considering the origin of reason, as gradually developed."[42]

One remarkable aspect of Darwin's notebooks is that they contain virtually no references to the sort of "complexity of design" argument so favored by Paley and all the natural theologians, including William Whewell at Cambridge, whose opinions Darwin greatly respected. In this he was true to his method, which was to amass evidence on species and generation first, rather than to start—as in the deductive mode— with a theoretical exercise. As he said in the *Autobiography,* "I worked on true Baconian principles, and without any theory collected facts on a wholesale scale."[43]

The months and years were now starting to pass quickly. He was still adding information but had closed in on his main idea linking heritable variation, population overproduction, the struggle for existence and, therefore, natural selection. There was huge satisfaction in having boiled down the dozens of overlapping and contradictory ideas and principles with which his predecessors had been occupied to a single, perfectly logical, thread.

As time went by, Darwin began to understand more fully, however, that there was an additional problem to solve. It would not be enough to propose a *theory* of evolution by natural selection, however logical; he would have to "prove" that it was correct. As he had written, "The whole object of the Work is its proof, its limiting, the allowing at same time true species. & its adaptation to classification & affinities, its extension."[44] He had fully to substantiate his theory. As he could not demonstrate unequivocally a case where one species had arisen from another (except for some cases of hybridization in plants), there was still a lot of work to do.

Meanwhile, having returned to London from his trip to Shrewsbury in July, Darwin had begun writing to Emma Wedgwood in very cozy terms. "Pray remember I consider myself invited to Maer, the next time I come down into the country.—in fact, I think I have been so often that I have a kind of vested right, so see me you will, & we will have another goose."[45] ("Goose" apparently meant "intimate talk.") On November 11, 1838, he visited Maer again, surprising everyone, including Emma, by popping the question. It was a Sunday; she said yes, and then left to teach her Sunday school class.

The wedding on January 29 was almost anticlimactic—unsurprisingly, neither the Lyells nor Horners were there, nor was Henslow. Darwin had a terrible headache. They had not planned a proper honeymoon, and in the end didn't take one. After the service, they traveled to London to a house Darwin had rented on Gower Street. The wedding feast was sandwiches eaten on the train, and they toasted each other with a bottle of water.

With capital of £10,000 settled on them by Dr. Darwin and £5,000 (and an income of £400 a year) from Emma's father, Darwin was financially secure. Five days before the wedding, he was elected fellow of the Royal Society. In the summer of 1839, the *Journal* was finally published, to great critical acclaim. Then, whatever he might have hoped his marriage to produce in terms of peaceful routine, yet another element was added to the tumult in December by the birth of a son, William Erasmus ("Hoddy Doddy").

EIGHTEEN

Finding His Place

If there is, as Shakespeare observed, "a tide in the affairs of men, which, taken at the flood, leads on to fortune," Darwin's tide had begun to flow. He was carried by it into a new world of ideas, scientific success, and personal anguish.

Between 1839 and 1842, while things began to slip into place for Darwin scientifically, his personal life was difficult. For all his connubial bliss, living in London had its ups and downs, with the downs becoming more and more oppressive. Neither he nor Emma really liked living in the city, with its appalling pollution from coal burning, the crowds, the smells and ordure from the ubiquitous horses, and the constant noise. They made many trips to Maer and Shrewsbury to get away. Darwin even came to suffer from the very aspect of life in the capital that made all the inconveniences worth it—the intellectual life. Living among the great scientific men of the capital (as well as all the social and political giants) both energized and debilitated him as, the more he refined his ideas of transmutation, the more he lived a lie.

Darwin's health failed to improve with marriage. If anything, it got worse. The main symptom was lassitude, often accompanied by vomiting and shivering. He constantly complained about his inability to work and became fixed on the idea that his condition was at least in part inherited and that he would never be among the strong, complaining, "What an unspeakable advantage they possess over us poor weak wretches."[1]

He repeatedly consulted his father, as well as many other doctors. The novelist Maria Edgeworth reported: "Dr Holland tells us that the voyage was not the cause, only the continuance of his suffering—for that before he went to sea he was subject to the same. His stomach rejects food continually; and the least agitation brings on the sickness directly so that he must be kept as quiet as it is possible and cannot see any body."[2] Darwin himself was well aware of a constant relationship between working too hard, stress, and his health.

The symptoms of Darwin's ill health had changed. By 1839, he was increasingly distressed by abdominal pain, flatulence, constipation, nausea, and headaches. Later, he started to have alarming episodes of inflammation of the whole face that were different from those of his youth. He also experienced joint pains and neural symptoms such as tingling in the fingers and legs, accompanied by fatigue and low fever. Some foods exacerbated the symptoms; others (plain, simple dishes) seemed to help.[3]

Most observers who have tried to diagnose Darwin's illness have tried to find a single cause; perhaps the most popular, over the years, has been the theory that he contracted Chagas' disease from a bug bite in Argentina in 1835.[4] The most recent interpretation, emphasizing the period of intense abdominal difficulty he experienced during the voyage (September and October 1834), is that Darwin's symptoms quite convincingly suggest Crohn's disease or another inflammatory bowel disease.[5] But neither of these would explain the facial eruptions of his youth (which seem to have had a different etiology than his later dermatological problems). It may be that Darwin suffered from two kinds of anxiety: one causing illness, the other the result of it. They then reinforced each other. In his youth, personal stress produced eczema, headaches, and palpitations; in his mature years, this was reinforced as a result of the (apparent) Crohn's disease. The extreme mental pressure created by the intensity of his working habits and his secrecy with respect to transmutation only made things much, much worse. The result was a negative spiral of health. Over a period of less than ten years, the dashing explorer of the *Beagle* voyage became an invalid with more than a touch of hypochondria.

Dr. Darwin prescribed calomel (mercurous chloride, presumably as a purgative) and was less than optimistic; Darwin reported: "My Father scarcely seems to expect, that I shall not become strong for some years—it has been a bitter mortification for me, to digest the conclusion, that the 'race is for the strong'— & that I shall probably do little more, but must be content to admire the strides others make in Science—So it must be, but I shall just crawl on with my S. American work & be as easy as I can."[6]

Being "easy" was the one thing that Darwin could not manage. Scientifically, he was operating at a high level. Despite his growing illness, he started his *Structure and Distribution of Coral Reefs* in October 1838, finishing it in 1842. Volumes dealing with fossil mammalia, birds, and fishes of the *Beagle* voyage were produced for the *Zoology*. He continued to work at geology subjects, publishing *Geological Observations on Volcanic Islands* in 1844 and *Geological Observations on South America* in 1846.

The Glen Roy project, on the other hand, turned into something of a disaster. In 1840, the young Swiss geologist Louis Agassiz published his idea (based in part on the work of Jean de Charpentier) that Europe had been covered relatively recently by a huge ice sheet. In November, he came to the Geological Society, where he showed that many previously difficult questions in geology—all those deposits of sand and gravel that had so concerned Sedgwick, the movement of erratic boulders, the carving out of valleys and, last but not least, phenomena like the parallel roads of Glen Roy—could be explained as the result of glacial action. The valley at Glen Roy had not been an arm of the sea but had been created by glaciers. The land had not been elevated. The "roads" were ancient beaches, but they had been formed at the edges of a succession of glacial lakes, dammed up in the valley. Darwin later groaned to Lyell, "My paper was one long gigantic blunder."[7] But he did not give up easily. He later wrote to Fox, "My marine theory for these roads was for a time knocked on the head by Agassiz ice-work—but it is now reviving again.—I dont mean, that I ever doubted, but others did (even Lyell for a time became a catastrophist) & they have now gone back to the elevation theory. The contrast between the valleys of

N. Wales, which have had all their rubbish & detritus swept out by the glaciers & those of Scotland, from which the sea has slowly retired, is very striking."[8]

With marriage to Emma, failing health, the arrival in rapid succession of William (1839) and Anne Elizabeth (1841), and with a third baby on the way (Mary Eleanor, born 1842, died after just three weeks), Darwin came to see that he needed to leave London. What he needed more even than exciting meetings of the Geological Society was access to the countryside: to walk, to ride, and to think. In the isolation of the countryside, there would be a blessed quiet in which to work. That would not be possible if they moved north to the Midlands where their families were. Shrewsbury was, in any case, too far from London, the importance of which in their lives could be diminished but not eliminated.

In 1842, Darwin found the ideal house in Kent. He lived for the rest of his life at Down House in the village of Down (later, Downe), a spot located sufficiently distant from the nearest railway station at Bromley to discourage casual visitors. The house came with eighteen acres of its own grounds and was set in glorious countryside. At first he had thought they would rent it; then Dr. Robert Darwin—with a better sense of the value of the investment—bought it (for £2,200).

Once they had moved to Down, Darwin settled into a new routine in which he was in a curious way both the head of the family and still a dependent. He has often been called a recluse, but in fact he was living totally on his own terms, choosing society or not as it pleased him but always avoided stressful occasions. He took part in the life of the local community. He even became a local magistrate. He kept fully in touch with colleagues and intellectual developments in London. As he labored over one project after another, it was a life of material ease but also of intense intellectual and physical discomfort.

In Emma he was blessed with a wife whom he came to adore, and who had an infinite capacity for concern over the most intimate details of his bodily functions and health. The household was soon organized with him as the central, supported figure. Emma and the servants looked after him, and the rhythms of the household were arranged entirely around him. They lived, as he said, a life like "clockwork."[9] Several times a day he would walk around the "sand walk" created in the garden where he did his best thinking. (In later years, there developed a regular routine: walk before breakfast, work, correspondence, Emma reading aloud, more work, another walk or ride, lunch, newspapers, letter writing, Emma reading aloud, work, dinner, more work after dinner.) He did not seem to work for more than an hour and a half at a time. He particularly enjoyed Emma's reading popular romances.

Darwin simply reveled in being a family man. It is hard to imagine the older Darwin without his family around him. The children were the source of his greatest joys and his very darkest, most devastating sorrows. The births and deaths of these children fall outside the scope of this book; there is no need to attempt here to rehearse what Randal Keynes has so vividly and tenderly written, for example, of the death of ten-year-old Annie in 1851.[10]

The image of Darwin as a happy, loving, and laid-back family man is difficult to reconcile with the intense, hard-driving, secretive, ambitious author of *On the Origin of Species* and the *Descent of Man.* Those were, in fact, two sides of the same coin; neither was possible without the other. Ambition and ill health were twin common factors, and also secrecy. In the period before the move to Down, Darwin kept his transmutationist thoughts very close to his chest for fear of being criticized or ridiculed. He was "anxious to avoid prejudice," as he put it in the *Autobiography.*[11] If he were to release his ideas too soon, ill formed and ill argued, he might lose altogether his chance to make a mark with them, or have them accepted at all. Even so, he would have had to be stoic beyond belief not at least to hint about what he was doing to close friends, perhaps hoping to strike a favorable response and encouragement from those he respected most.

Not surprisingly, it may have been Fox to whom he opened up first. Just before his trip to Glen Roy (and Maer!), he told Fox that "the crossing of animals . . . is my prime hobby & I really think some day, I shall

5. Darwin and his firstborn, William, aged eighteen months, 1842 daguerreotype. (Courtesy of Cambridge University Library [DAR 225.129].)

be able to do something on that most intricate subject species & varieties."[12]

Lyell was perhaps the toughest audience for any thought of transmutation (except Sedgwick, whom Darwin did not try at all). He broached the subject with Lyell in the most innocuous terms on September 14,

1838. "I have lately been tempted to be idle (with respect to geology) . . . by the delightful number of new views, which have been coming in, thickly & steadily, on the classification & affinities & instincts of animals— bearing on the question of species—note book, after note book has been filled, with facts, which began to group themselves clearly under sub laws."[13] By most reckonings of Darwinian chronology, this was just before he read Malthus again, and the language is carefully couched so that Lyell might think that by the "question of species" Darwin was referring to their constancy, rather than transmutation.

Henslow was not going to be an easy convert to transmutation, either. It was another year before Darwin wrote to him about "steadily collecting every sort of fact, which may throw light on the origin & variation of species."[14]

All Darwin's reading and thinking constantly led him to new questions, most often in the area of plant and animal breeding, and the heritability of variation. Sometime in 1839, he opened a notebook marked "Questions & Experiments" in which he recorded thoughts and observations in these areas. It was clear that the creation of new varieties and races through artificial selection (breeding) was a central metaphor for what he would soon term natural selection. He had started a massive, lifelong correspondence with experts in various subjects—from breeding fancy pigeons to the sex lives of tropical plants, from the survival of seeds in salt water to the instincts of mammals. Where possible, he conducted his own experiments in the same subjects.

Naturally, Fox was one of his main correspondents. In a letter that captures the flavor of this effort, Darwin wrote: "I continue to collect all kinds of facts, about 'Varieties & Species' for my some-day work to be so entitled—the smallest contributions, thankfully accepted— descriptions of offspring of all crosses between all domestic birds and animals dogs, cats &c &c very valuable—Don't forget, if your half-bred African Cat should die, that I should be very much obliged, for its carcase sent up in a little hamper for skeleton.—it or any cross-bred pidgeon, fowl, duck, &c &c will be more acceptable than the finest haunch of Venison or the finest turtle. . . . should you ever have opportunity when in Derbyshire, do enquire for me, from some person you told me of whether offspring of male muscovy & female common duck, resembles offspring of female muscuvy & male common—How many hybrid eggs are produced."[15]

This process of asking questions and collecting data became a lifelong pattern of working, applied to every subject that Darwin attacked. With respect to evolution, it was a necessity because the subject was never fully closed. Darwin could never answer all the questions. The range of his probing intellect and imagination was extraordinary. A century and a half later, modern evolutionary biologists find that there are precious few aspects of evolutionary biology that he had not inquired into. Modern ecology, genetics, and molecular biology follow in the paths he first broke.

NINETEEN

First Drafts

Darwin had intended all along to develop his species theory into a full-length book. It was not until 1842, however, that he attempted to write even the briefest summary of what that book would include (at least insofar as his records have been preserved). In May, he took the family first to Maer and then to Shrewsbury. Sometime in May or June, he sat down and wrote out his theory, first as three chapter headings, then as fifty-three pages of crabbed notes written, as his son Francis Darwin was later to say, "on bad paper with a soft pencil, and . . . in many parts extremely difficult to read . . . written very rapidly."[1] This *Sketch* was more like an extension of the notebooks rather than a full text. "The whole is more like hasty memoranda of what was clear to himself, than material for the convincing of others."[2]

The critical issue for Darwin was to develop a strategy for making his case. As he had written four years earlier in Notebook C, "It will be easy to prove persistent Varieties in wild animals—but how to show species—I fear argument must rest on analogy . . . it may be said argument will explain very close species in islds. Near continent. Must we resort to quite different origin when species rather further . . . once grant good species . . . & analogy will necessarily explain the rest."[3]

Darwin first thought of three chapters: "I. The Principles of Var. in domestic organisms; II. The possible and probable application of these same principles to wild animals and consequently the possible and probable production of wild races, analogous to the domestic ones of plants and animals; III. The reasons for and against believing that such races have really been produced, forming what are called species." When he started writing it all out, the chapters increased to ten, and these very closely foreshadow the arrangement of subjects in his *On the Origin of Species* of 1859.

To persuade what he expected to be a mostly skeptical audience, he chose the (Baconian and Newtonian) approach of vera causa that Herschel had explicated in his *Preliminary Discourse on Natural Philosophy*. He also borrowed the tactics of William Paley, who had opened his *Natural Theology* with a key analogy—the watch that must have had a designer. Darwin opened with a chapter on variation under artificial selection—a process that demonstrably produces change and new varieties and races. This established artificial selection as the model for natural selection.

Darwin's second chapter made the case that animals and plants vary naturally and presented the Malthusian basis of the struggle for existence. Logically, then, this produced natural selection. Natural selection was therefore set up as the vera causa of evolution. Vera causa requires not only that a mechanism exists but that it be capable of producing the effects under investigation. The remaining chapters were intended to provide just that proof—in Darwin's favorite fields, geology and biogeography. Then Darwin showed how "unity of type" is explained by evolution and adaptation, finally discussing "Abortive Organs" such as the presence of tiny internal limb bones in some snakes. Then there remained only the "Recapitulation and Conclusion."

To select a few highlights: the first chapter showed both the strength and the weakness of the analogy of artificial selection and the nature of variation under domestication. Its third paragraph reads (in part): "When the organism is bred for several generations under new or varying conditions, the variation is greater in amount and endless in kind. . . . The nature of the external conditions tends to effect some definite change in all or greater part of offspring . . . simple generation especially under new conditions [causes] infinite variation." Darwin was not sure whether or how the environment, acting directly on the phenotype, could produce changes that were heritable. He had long since rejected Lamarck's ideas on use and disuse, but traces of the concept of the inheritance of acquired characters persisted. These were carried into the second, much longer draft that Darwin wrote in 1844, and even into the *Origin* itself.

The second chapter was strong in making the case for the mathematical basis of natural selection but a little weak in its demonstration of variation in nature. "Wild animals vary exceedingly little. . . . Nature's variation far less, but selection far more rigid and scrutinizing." Darwin's argument had to depend on his statement that "any and every one of these organisms would vary if taken away and placed under new conditions. Geology proclaims a constant round of change, bringing into play, by every possible change of climate and the death of pre-existing inhabitants, endless variations of new conditions."

Here Darwin was extrapolating directly from the sort of variation that Henslow (among many others) found when seeds were planted in highly manured test beds. And he followed with a reference to Buffon: "According to nature of new conditions, so we might expect all or majority of organisms born under them to vary in some definite way. Further we might expect that the mould in which they are cast would likewise vary in some small degree."

In this chapter, it became clear that Darwin had all along been developing a subtheory of his natural selection. Separate from the competition for food, space, and the like, in many species there was intense and different selection in males versus females. The result was the development of elaborate secondary sexual characteristics such as the tail in the male peacock or the red in the male woodpecker's neck or the mane of the lion. (Darwin's concept of "sexual selection" helped encourage later psychological theorists to imagine that Darwin was somehow unhealthily preoccupied by sex.)

The chapter ended with a frank discussion of "difficulties on theory of selection." Here he faced the challenge of explaining the origin of major organs such as the eye or ear. "But think of the gradation, even now manifest. . . . Everyone will allow if every fossil preserved, gradation infinitely more perfect . . . every analogy renders it probable that intermediate forms have existed; part of the eye, not directly connected with vision, might come to be [thus used] . . . swimming bladder by gradation of structure is admitted to belong to the ear system."

He concluded that "the gradations by which each individual organ has arrived at its present state, and each individual animal with its aggregate of organs, probably never could be known, and all present great difficulties. I merely wish to show that the proposition is not so monstrous as it at first appears, and that if good reason can be advanced for believing the species have descended from common parents, the difficulty of imagining intermediate forms of structure not sufficient to make one at once reject the theory."

The second part of the *Sketch* sought to answer the question posed at the very end of the first part: "Is there any evidence that species have been thus produced, this is a question wholly independent of all previous points, and which on examination of the kingdom of nature ought to answer one way or another." In the geological section, Darwin relied on the progressive nature of the fossil record but noted the imperfection of the geological record and frankly admitted that, if theories like those of Lyell on the cyclic nature of earth history were true, "my theory must be given up." He noted that extinction remained a problem, but: "The first fact geology proclaims is immense number of extinct forms, and new appearances . . . forms gradually become rare and disappear and are gradually replaced by others . . . we know of no sudden creations."

The following chapter, on geography, took as its theme the changes that geology produces in geography, leading to changes in animal and plant distribution, to isolation of populations, and to changing environmental conditions. The biogeographical-geological argument was continued in the following chapter, "Affinities and Classification." For example: "As all Mammals have descended from one stock, we ought to expect that every continent has been at some time connected, hence obliteration of present ranges." And, while descent from common ancestry explains *homology*, "external conditions air, earth, water being same on globe . . . organisms of widely different descent might become adapted to the same end and then we should have cases of *analogy.*"

The next chapter explained "Unity of Type" in terms of evolutionary homology. Here he finally chose sides in the debate between Meckel and von Baer over the significance of the fetal stages of animals. He did not accept the recapitulation argument. "[The] general unity of type in great groups of organisms . . . displays itself in a most striking manner in the stages through which the foetus passes. . . . It is not true that one passes through the form of a lower group."

In his section on abortive organs, he continued the argument with a summary of the power of his theory. "If abortive organs are a trace preserved by hereditary tendency, of organ in ancestor of use, we can at once see why important in natural classification, also why more plain in young animal because, as in last section, the selection has altered the old animal most. . . . part created for no use in past and present time, all can by my theory receive simple explanation; or they receive none and we must be content with some . . . empty metaphor."

The *Sketch* of 1842 is not quite the complete theory as it has come to be known to us today after six editions of the *Origin* and 150 years of amplification. What strikes the reader most forcibly is the way in which Darwin was hampered by the lack of any science of genetics. This accounts for what we might now, all-knowingly, call error, while at the same time it points up his brilliance in having created his theory without such a foundation. Without genetics, the basis for variation among individual organisms was missing, and the very first sentences of the *Sketch* showed that Darwin was still dependent on ideas derived from Lamarck concerning the origin of variation.

The *Sketch* of 1842 contains no argument from the design principle of Paley's *Natural Theology,* but Darwin knew he had to deal with that subject. In the conclusion he wrote: "We no longer look on animal as a savage does at a ship, or other great work of art, as a thing wholly beyond comprehension, but we feel far more interest in examining it." This sentence is preserved almost verbatim in the *Origin* of 1859 (he took out the reference to a "work of art").[4]

Darwin's last pages summed up perfectly his new approach to nature and his rejection of the old. Perhaps a little defensively, he used the word "grandeur" twice. The language used was polished, quite in contrast to the rest of the *Sketch*. It was obviously a section on which he labored long and lovingly. The result is Darwin's writing at its very best.

> There is much grandeur in looking at existing animals as either the lineal descendents of the forms buried under thousand feet of matter, or as the coheirs of some still more ancient ancestor. It accords with what we know of the law impressed on matter by the Creator, that the creation and extinction of forms, like the birth and death of individuals should be the effect of secondary means. It is derogatory that the Creator of countless systems of worlds should have created each of the myriads of creeping parasites and slimy worms which have swarmed each day of life on land. . . . We cease being astonished, however much we may deplore, that a group of animals should have been directly created to lay their eggs in bowels and flesh of other,—that some animals should delight in cruelty, . . . —that annually there should be an incalculable waste of eggs and pollen. From death, famine, rapine, and the concealed war of nature we can see that the highest good, which we can conceive, the creation of the higher animals has come. . . . There is a simple grandeur in this view of life with its powers of growth, assimilation and reproduction, being originally breathed into matter under one or a few forms, and that whilst this our planet has gone circling on according to fixed laws, . . . that from so simple an origin, through the process of gradual selection of infinitesimal changes, endless forms most beautiful and most wonderful have evolved.

This elegant passage runs the gamut of positions from deist ("the effect of secondary means") to antireligious ("derogatory that the Creator") to opposing natural theology ("directly created to lay their eggs in bowels") to an endorsement of Malthus ("the concealed war of nature") and finally to his own theory ("from so simple an origin . . . evolved").

Having written out this outline of a future book, Darwin had crossed a threshold. He saw that he could indeed achieve something worth publishing. And at Down House he had the tranquility and independence within which to work. He still continued to create several books and papers at once, but transmutation began to occupy more and more of his time, and he started to try out his idea on a few more people.

Correspondence over the plants of the Galapagos (the *Beagle* collections still had not been worked up) led Darwin to meet the young botanist Joseph Dalton Hooker. Hooker was the son of William Jackson Hooker, Henslow's great friend and first director of the Royal Botanic Gardens at Kew. Joseph Hooker had had his own foreign adventures in the service of science. He had been a member of James Ross's expedition to the Antarctic (1838–43) and later made an expedition to the Himalayas, collecting plants. And although Darwin did not know Hooker as well as he knew Henslow, Lyell, Jenyns or, certainly, Fox, it was Hooker to whom Darwin next ventured to reveal a little more of the dangerous thoughts he had been harboring. Darwin's letter is worth quoting at length.

> I have been now . . . engaged in a very presumptuous work & which I know no one individual who wd not say a very foolish one.—I was so struck with distribution of Galapagos organisms &c &c & with the character of the American fossil mammifers, &c &c that I determined to collect blindly every sort of fact, which cd bear any way on what are species.—I have read heaps of agricultural & horticultural books, & have never ceased collecting facts—At last gleams of light have come, & I am almost convinced (quite contrary to opinion I started with) that species are not (it is like confessing a murder) immutable. Heaven forfend me from Lamarck nonsense of a "tendency to progression" "adaptations from the slow willing

of animals" &c,—but the conclusions I am led to are not widely different from his—though the means of change are wholly so— I think I have found out (here's presumption!) the simple way by which species become exquisitely adapted to various ends.— You will now groan, & think to yourself "on what a man have I been wasting my time in writing to."—I shd, five years ago, have thought so.—I fear you will also groan at the length of this letter—excuse me, I did not begin with malice prepense.[5]

That Hooker was neither antagonistic to Darwin's idea nor even very surprised shows how much the concept of mutability of species had started to take hold, at least among some of the younger generation of naturalists to whom Lamarck's ideas were not complete nonsense. Their relationship began to flourish, with Hooker eventually becoming one of Darwin's most trusted confidants.

As Darwin's phrase "like confessing a murder" suggests, his chronic anxiety persisted and, even though he and his growing family were co-cooned at Down House, his health inevitably and dramatically worsened as his evolutionary ideas firmed up. The pressure on him came not just from the fear of ridicule by the scientific establishment but from his likely reception by the religious world as well. Transmutation was not merely a scientific idea; it had profound theological implications about the literal truth of the Bible and the role of God in worldly affairs. And the closest person to him who was really religious was his own wife, Emma. Darwin wrote in the *Autobiography* about how his views of religious belief changed from those of the conventional Christian (during the *Beagle* voyage and before) to agnosticism to a form of atheism. The personal anguish this caused them both was considerable, and Emma redacted those sections of the *Autobiography* from the first published edition.

Darwin wrote out a much longer draft of his theory in 1844. This 1844 *Essay* was closer to being a finished work, and Darwin wrote Emma an "In case of my death" letter that he hoped would ensure its publication. This in itself presented some difficult choices—whom should he nominate to see it through? "Mr Lyell would be the best . . . I believe he wd find the work pleasant & he wd learn some facts new to him." After that the choices were harder, although "quite the best in many respects would be Professor Henslow. . . . Professor Owen wd be very good, but I presume he wd not undertake such a work."[6]

In the reorganization of topics, the most striking difference between the 1844 *Essay* and the *Origin* of 1859 was that the discussion of instinct had been moved (and in fact could easily have been omitted, except that this was one of Darwin's favorite topics). The first three chapters from the 1842 and 1844 versions became an exposition of the entire theory. Discussions of classification, affinities, embryology, abortive organs, and so on were condensed to one chapter. A chapter was added on geological succession.

Of the various parts of Darwin's theory, the struggle for existence and the potential for natural selection were, and are, incontestable. The weakness still lay with variation. In our modern view, selection can act only on variation that is thrown up by the constant churning of the genetic system due to crossing (Darwin's "generation"). But, as the causes of variation were then unknown, Darwin was forced either to fall back on a Lamarckian inheritance of acquired characteristics or to invoke a "black-box" mechanism that he could not demonstrate. Added to this was the question (still pertinent today) of whether the sorts of minor differences observed in the variation within living populations of animals and plants were sufficient in scale to accumulate, even ever so slowly, to produce new organs like the eye or ear. This is a problem that has not been completely solved today.

There has been much speculation on the fascinating question of why Darwin waited from 1844 to 1859 to publish his theory of evolution by natural selection. The most obvious answer was that there was more work to be done. The 1844 *Essay* was, in one sense, almost a finished work, and Darwin saw that there was enough for someone else to publish if he should happen to die. But Darwin knew that it might not be enough to make his case—to carry the day against the antievolutionists. His whole idea could only be presented as "one long argument" rather than a proven, demonstrated fact. This was exactly the same dilemma that had faced Paley in *Natural Theology* and, like Paley, Darwin attacked it by piling on more information. He stepped up his massive correspondence with authorities around the world concerning every aspect of his theory—artificial selection (plant and animal breeding), variation, superfecundity, the geological record, biogeography, and embryology. Where possible, he made direct experiments himself—for example, with fancy strains of pigeons—something that harked right back to his childhood days with his mother at the Mount. He read every possible pertinent book. And he delved more deeply than he had before into the metaphysical aspects of the subject. There was no telling when enough would be enough, and Darwin was in no hurry to publish too soon.

Darwin knew that the absence of concrete proof would always be a difficulty for his theory. One cannot, for example, rerun history to demonstrate the evolution of the eye; one can only demonstrate the existence in living organisms of every graded sort of structure, starting with a simple photosensitive cell, and conclude that the eye evolved. Just like Paley, therefore, what Darwin had to do was to craft a work with overwhelming documentation for every tiny element in his theory.

In 1844, there came the one event for which he had not planned and after which it no longer mattered whether he intended to delay publication or for how long: an anonymous author apparently scooped him. In Edinburgh, Robert Chambers, previously known as an author of historical works and as the publisher of *Chamber's Cyclopedia* and other reference works, published (anonymously) his *Vestiges of the Natural History of Creation*. This was an exciting, up-to-date presentation of a view of evolution with hints of Lamarckism and built on the concept of the progressive nature of the fossil record. Chambers tried to develop a theory not just of living organisms but also the whole cosmos. He did not hesitate to link humans with the lower animals. *Vestiges* quickly became as popular with the public as it was excoriated by scientists and clergy.

Vestiges is a curious mixture of a book. The first edition was full of scientific errors, but it had one strong argument: species evolved. Chambers had proceeded roughly as Darwin had, by collecting all the known data and gleaning from them a viable theory. The presentation was dramatic but, in retrospect, there were lots of problems with his facts and his logic. Darwin read it and was evidently impressed, although he thought Chambers had his geology all wrong. Over the next few years, as new, improved editions of *Vestiges* appeared, Darwin became certain that he could make a far better case with his natural selection.

The whole scientific establishment came out and savaged *Vestiges,* which scared Darwin—would they then do the same to him? Sedgwick's review was so hostile that even some fellow scientists were a bit embarrassed. But on the other hand, *Vestiges* was something of a stalking horse for Darwin. The attacks showed Darwin just the sort of criticisms he must be prepared to encounter and *counter.*

Whether or not Chambers had made his point scientifically, Darwin learned two valuable lessons: the subject was still just as controversial as he had feared, and his scientific colleagues were not prepared to give transmutation—in any shape or form—any consideration. No doubt sober reflection told him that the scale of the antagonism revealed just how shaky was the scientific establishment's basis for its stand against transmutation. That establishment, however, was what would make or break his own work. Faced with that, Darwin had to do two things: he had to make his arguments as foolproof and Sedgwickproof as possible, and he had to wait—both for the dust to settle and for his own work to be made complete.

The delays caused by Chambers were, on balance, distinctly to Darwin's advantage. When MacLeay had created his Quinary System, he did so on the basis of a detailed analysis of the relationship of a single large family of beetles, the scarabs. Darwin now sat down in his study at Down and began what would turn out to be an eight-year study of the structure, reproduction, life histories, and classification of Cirripedia— the barnacles. Barnacles had always fascinated him, and no one had been found to work up his large *Beagle* collections of that group. Harking back to his days with Grant in Edinburgh, studying the minute details of the anatomy of lower animals, in dogged fashion Darwin turned to the microscope and produced, in 1851 and 1854, a two-volume monograph, *Living Cirripedia,* that stands today as a classic in biology.

Darwin's *Cirripedia,* which (along with his geological work) won him the Royal Medal of the Royal Society in 1853, produced several important results for the theory that he was still amplifying and polishing quietly. Here, in one minutely studied group, were adaptation, homology, convergence, and variation laid out in extraordinary fullness. *Cirripedia* established beyond question his bona fides as a zoologist as well as a geologist.

There turned out to be yet another advantage to the delay, because a key element of his theory had not yet occurred to him. When it did, he was suitably chagrined. He had "overlooked one problem of great importance . . . the tendency of organic beings descended from the same stock to diverge in character as they become modified."[7] If "descent with modification" were to occur only in straight lines of descent, then there would be no divergence. However, if descent is coupled with branching, divergence is inevitable.

And still Darwin did not publish his transmutation theory. He kept plugging away at the details, and he was no less assiduous in cultivating an audience for his ideas by carefully revealing them to more and more key people and trying to persuade them to his side.

Then, as the world knows, lightning struck again. In June 1858, fourteen years after Chambers's book had been first published, a naturalist with whom Darwin had been in correspondence for many years (exchanging information about animal and plant distributions) sent a package from Indonesia. Alfred Russel Wallace was an explorer and naturalist; in many ways, he was the sort of non–gentleman naturalist, a professional collector, for whom Darwin had not wanted to be mistaken when being considered for the *Beagle* voyage. He traveled and collected widely in South America and the East Indies and, from his reading of Lamarck and Geoffroy, was well disposed to the principle of transmutation. Wallace was a man of formidable accomplishment. His 1869 book *The Malay Archipelago* became one of the great classics of natural history, and he very much surpassed Darwin as the master biogeographer of the nineteenth century. Any communication from Wallace was going to be worth reading.

TWENTY

Crisis and Resolution

The package that Wallace sent from the island of Ternate (or possibly Gilolo) in Indonesia contained a manuscript that he wanted Darwin to submit on his behalf to the Linnean Society of London. *On the Tendency of Varieties to Depart Indefinitely from the Original Type* was nothing less than a précis of the theory of natural selection. Wallace had had the same idea as Darwin, and had even been triggered into it by reading Malthus and observing, face-to-face, the struggle for existence that goes on daily and hourly in nature.

Darwin was devastated—although, in fact, more surprised than he might have been if he had correctly interpreted a paper, "On the Law Which Has Regulated the Introduction of Species," that Wallace had published in 1855. Darwin, Lyell and others had not fully grasped where Wallace was heading. It was Darwin's worst nightmare come true. Surely he must long have worried that, by delaying publication, he was increasing the chance that someone else might really scoop him; someone might come up with something much closer to home than Chambers's ideas. After all, right at the beginning of his career, thirty years previously, Grant had "stolen" his observations on flustra and *Pontobdella*. Once bitten, twice shy—as the *Autobiography* shows, Darwin did not easily share the credit for anything. Was he now to lose priority for his great idea of natural selection—essentially his life's work? Aware of the danger, in 1856, Lyell had urged him to write up his ideas more fully. Darwin replied, somewhat coyly: "I rather hate the idea of writing for priority, yet I certainly should be vexed if anyone were to publish my doctrine before me."[1] Now his "priority" was in serious danger and, in fact, lost. From this point, the theory of evolution by natural selection became the Darwin-Wallace theory, although the force of Darwin's personality, the growing momentum of his career, and the number of his friends and supporters created a juggernaut that would forever give by far the greater share of credit to Darwin. Indeed, while Wallace had written out the bare outlines of the idea of natural selection, Darwin had an overwhelming mass of facts with which to support it. And, by taking the credit, Darwin would also absorb the lion's share of the criticism.

Whether or not (probably not) Darwin at the time put on the face of good sportsmanship and resignation that he later expressed in his *Autobiography*—"I cared very little whether men attributed most originality to me or Wallace"—Darwin and his supporters hurriedly went to work to rescue the position.[2] Darwin's careful cultivation of influential scientists, followed with great vigor after the Chambers's affair, now paid dividends. In quietly canvassing for his theory over the past few years, Darwin had built up a group of supporters who knew that, despite his failure to publish, he had priority for the idea. One of those was the American botanist Asa Gray at Harvard. Darwin had sent him an outline of his theory in 1857.[3] Lyell and Hooker proposed an honorable (even Solomonic) compromise. Darwin would prepare a short paper on his own work and it would be read side by side with Wallace's at the Linnean Society in London, along with Darwin's letter to Gray and an extract from his manuscript. After that, Darwin should proceed at full speed with his book.

Yet another statement in the *Autobiography* seems scarcely believable: "I was at first very unwilling to consent, for I did not then know how generous and noble was his [Wallace's] disposition."[4] It defies human nature for either of them to have been *that* totally self-effacing, although history shows that Wallace was indeed content (or resigned) to have Darwin take the major credit. In this respect, Darwin was extremely fortunate that it was Wallace, and not someone more formidably combative, like Richard Owen, whose ideas had converged with his. In later years, Wallace earned his own fame with his masterpiece *The*

Malay Archipelago, and lost some of it by embracing the fashionable Spiritualism of the next decade.

The two papers were read on July 1, 1858, to surprisingly little debate.[5] It was all something of an anticlimax. And then, the threat having been dealt with surprisingly easily, Darwin set to work in earnest. By now, the working version of his projected work had become mammoth in size, bloated with minute details. In one sense, posterity was blessed in that Darwin had to sit down and write out a "mere abstract" of some 490 pages, entitled *On the Origin of Species by Means of Natural Selection, or the Preservation of Favoured Races in the Struggle for Life.*

In the book, Darwin followed the same sequence of argument as in the *Sketch* and *Essay.* The *Origin* falls into three sections. First, he established artificial selection as a vera causa of change in animal and plant breeding; then he established the phenomenon of variation in nature; this was followed by a demonstration of the struggle for existence; next he revisited variation and ended this, the fifth, chapter with a summary of his theory of natural selection. The second section started with a frank discussion of "difficulties on theory" that both admits those difficulties and solves them. There were two chapters on instinct and hybridism; the imperfections of the geological record; geological succession of types (harking back to Lyell and Sedgwick from his Cambridge days, and, of course, Lyell's *Principles*). The third section was made up of geographical distribution, classification, affinity, homology, and embryology. Then, finally, his recapitulation and conclusion.

The first edition was of a mere 1,250 copies. It sold out on the first day, which suggests not only that Darwin's was an idea whose time had come, but that Darwin's cultivation of his colleagues had paid off. Sedgwick was not persuaded, but overall, instead of being ridiculed, Darwin's theory was taken seriously. By most accounts, the origin of *The Origin* had been a journey of twenty-two years; really, it had taken a lifetime.

Our story ends here, at the moment that must be considered the dawn of modern biological science. With *On the Origin of Species* and the

Darwin-Wallace concept of natural selection, Darwin had given the life sciences a fundamental theory and in the process had created a clear intellectual link to the earth sciences. The theory would not be complete (on his terms at least) until he had added a volume, *The Descent of Man,* because he had strategically not burdened *On the Origin* by entering into the even more dangerous territory of human origins. But even that most sensitive subject had been at the back of the minds of many scholars for years, if not generations.

Any biologist studying evolution today quickly discovers that there is very little *conceptual* in the subject that had not been thought about first by Darwin. This is the more remarkable because he worked without a knowledge of molecules, genes, and the mechanisms of development—all those factors that we see as the mechanistic nuts and bolts of biology. His tripartite theory—variation, superfecundity, selection—has been amplified in many areas, but never supplanted.

Darwin himself trod familiar ground when he first decided, in the spring of 1837, to investigate "transformism." What we call evolution had long since been mooted by others. What Darwin brought to the subject was a deep grounding in practical natural history. Always looking for logical systems and rational patterns of knowledge, he was already an expert on insects while still a student at Cambridge. To this, during the critical years of the *Beagle* voyage, he added both a strong foundation in scientific method, as developed in the study and practice of geology, and a keen sense of the patterns of order and disorder represented by the geographical distributions of animals and plants. Others might have had the same experience and produced different, lesser results. The final ingredient that Darwin brought to the study of transmutation of species was a particular mindset—an acute intelligence, a dogged perseverance, and the ability to synthesize information and ideas from a wide range of subjects.

Like many natural scientists, Darwin was a keen collector as a child; even more significant, he was always someone who wanted to understand things, not simply document them. Intense, self-absorbed, brilliant, an avid reader since childhood, intellectually independent (even though emotionally dependent on others): in retrospect, it seems quite impossible that Darwin would have ever fit comfortably into any of the conventional professions, even though life as a doctor or clergyman would have afforded him plenty of opportunities for scholarship. The world owes Dr. Robert Darwin and the Wedgwood family a huge debt for the financial freedom they afforded Darwin.

The very breadth of Darwin's learning, his catholic interests in natural science, and his passion for both the practical and the philosophical meant that his path to his theory of natural selection was tortuous (it cannot be considered slow, as he accomplished so much in a five-year period while mostly working on other subjects). His goal was not to discover some narrow set of facts concerning the transmutation of species but to articulate an entirely "New System of Nature," in which geology, biogeography, formal structure and physiological function, embryology, heredity, mind and instinct were brought together. That meant that he had to explore the contemporary state of each of these fields as well as to sift and digest the theories of evolutionary change proposed by Buffon, Lamarck, and Erasmus Darwin; the systems of formal anatomy being developed in Germany and France; and also purely classificatory systems such as MacLeay's Quinary System. Everything had to be weighed in the logical balance, the valuable parts of each retained and the dross rejected. The process took him down many blind alleys.

In following the progress of Darwin's ideas, as expressed in his scientific notebooks, we see a very human Darwin who sometimes found it hard to give up a favorite notion. We see a certain ruthlessness when it came to the ownership of ideas. We also see the sheer brilliance with which he was able to seize upon a crucial element and make an intellectual breakthrough in the association of facts and theories.

It must have been very hard for Darwin's family and friends to live with him in the years between 1837 and 1844. Single-mindedness is much more palatable when seen from afar. Although his health was a tremendous trial for him and for his family and friends, it has also to be seen as part and parcel of the man and of the creative process. Illness was caused by the demands of his personality and the pressure of his work; it also contributed positively to his creative self-absorption.

In the end, Darwin, like so many great men, has to be seen as a series of contradictions, not all of which we will ever fully understand. He
was variously a dashing, daring explorer and a recluse, a competitive,
driven scholar and a loving, gentle family man, a student of minutiae
and a grand theorist, a revolutionary and the founder of an orthodoxy.
A man who avoided controversy whenever possible, Darwin might be
surprised at the overheated rhetoric of modern debates over science
and religion—but then, given his sensitivity, right from the beginning,
to the religious implications of his ideas, he might not. Perhaps it is only
fitting that Darwin, an atheist (or at least an agnostic) but a man who
tried to bring things together rather than to divide, was buried in
Britain's pantheon, St. Paul's Cathedral.

Notes

Quotations from Darwin's letters are taken from the definitive *The Correspondence of Charles Darwin,* ed. Frederick Burkhardt and Sydney Smith (Cambridge: Cambridge University Press, 1985–); also available online at www.darwinproject.ac.uk.

Quotations from Darwin's scientific notebooks are cited by the individual notebook and, unless otherwise specified, are from *Charles Darwin's Notebooks, 1836–1844,* ed. Paul. H. Barrett, Peter J. Gautrey, Sandra Herbert, David Kohn, and Sydney Smith (Ithaca, N.Y.: Cornell University Press, 1987).

Quotations from Darwin's two autobiographical essays are given as *Autobiography* and taken from *Charles Darwin: Autobiographies,* ed. Michael Neve (London: Penguin Classics, 1986).

Quotations from Darwin's diary of the voyage of the *Beagle* are given as *Beagle Diary* and taken from *Charles Darwin's Diary of the Voyage of H.M.S. Beagle,* ed. Nora Barlow (Cambridge: Cambridge University Press, 1934).

ONE
Falmouth

1. *Beagle Diary,* 431.

2. Caroline Darwin to Sarah Wedgwood, October 5, 1836, in *Correspondence,* 1:304–5.

3. Darwin to Robert FitzRoy, October 6, 1836, in *Correspondence,* 1:506–7. "Philos"—for Philosopher—was FitzRoy's nickname for Darwin.

4. B. J. Sulivan, *The Life and Letters of the Late Admiral Sir B. J. Sulivan, KCB, 1810–1890,* ed H. N. Sulivan (London: Murray, 1896), 46.

TWO
Antecedents

1. *Autobiography,* 4

2. Desmond King-Hele, *Erasmus Darwin* (London: Giles de la Mare, 1999).

3. Charles Darwin, in Ernst Krause, *Erasmus Darwin* (New York: Appleton, 1880), 45.

4. Jenny Uglow, *The Lunar Men* (London: Faber and Faber, 2002).

5. Christopher U. M. Smith and Robert Arnott, eds., *The Genius of Erasmus Darwin* (Burlington: Ashgate, 2005).

6. Erasmus Darwin, *Zoonomia; or, The Laws of Organic Life* (London: Johnson, 1794), vii.

7. Ibid., 395.

8. Ibid., section 29, "Of Generation."

9. Darwin, in Krause, *Erasmus Darwin*, 27.

10. Anna Seward, *Memoirs of the Life of Dr. Darwin, Chiefly during His Residence in Lichfield, with Anecdotes of His Friends, and Criticisms on His Writings* (London: Johnson, 1804).

11. In later life, Darwin tried to recoup the reputation of his grandfather in his preface to Krause's biography of Erasmus Darwin.

12. Seward, *Memoirs*, 297.

13. John Bowlby, *Charles Darwin: A New Biography* (London: Hutchinson, 1990), 44.

14. Frances Anne Violette Darwin married Samuel Tertius Galton, and among her sons was Darwin's cousin Francis Galton. Harriet Darwin married an admiral. Emma Georgina Darwin became a teacher. Of the sons of Erasmus Darwin who survived childhood, John became a parson and Edward entered the army. Francis Sachervell Darwin (later Sir Francis) and Robert Waring Darwin became doctors like their father. Susan and Mary Parker, Erasmus Darwin's illegitimate daughters, both became schoolteachers.

15. Darwin, in Krause, *Erasmus Darwin*, 86.

16. Bowlby, *Charles Darwin*, 41.

17. *Autobiography*, 18.

18. Eliza Meteyard, quoted in Bowlby, *Charles Darwin*, 46.

19. Ibid.

20. Susannah Darwin to her brother Josiah Wedgwood, quoted in Bowlby, *Charles Darwin*, 53.

THREE
Childhood

1. *Autobiography*, 3, 7.

2. Ibid., 10.

3. My friend Donald Cresswell has shown me the copy of the Lewis and Clark book that circulated in a private lending library in Shrewsbury, with Dr. Darwin's name as a reader.

4. *Autobiography*, 21.

5. Quoted in S. M. Walters and E. A. Stow, *Darwin's Mentor, John Stevens Henslow, 1796–1861* (Cambridge: Cambridge University Press, 2001), 84.

6. *Autobiography*, 21–22.

7. Ibid., 7.

8. Ibid.

9. Ibid., 2–3.

10. By far the best discussion of Darwin's life in this regard is by the child psychiatrist Dr. John Bowlby: *Charles Darwin: A New Biography* (London: Hutchinson, 1990).

11. Ibid., 62.

12. Janet Browne, *Charles Darwin, Voyaging* (New York: Knopf, 1995), 20.

13. Ralph Culp, *To Be an Invalid: The Illness of Charles Darwin* (Chicago: University of Chicago Press, 1977), 6.

14. Darwin to Fox, January 25–29, 1829, in *Correspondence*, 1:73–74.

15. Edward Kempf, *Psychopathology* (New York: Mosby, 1931), 209.

16. Francis Darwin, ed., *Life and Letters of Charles Darwin* (London: Murray, 1887), 1:26; Kempf, *Pyschopathology*, 213.

17. *Autobiography*, 23.

FOUR
Edinburgh

1. Erasmus to Darwin, February 24, 1825, in *Correspondence*, 1:15–16.

2. Darwin to his father, October 23, 1825, in *Correspondence*, 1:19.

3. Ibid.

4. George Shepperson, "The Intellectual Background of Charles Darwin's Student Years at Edinburgh," in *Darwinism and Society,* ed. Michael Banton (Chicago, Quadrangle, 1961).

5. *Autobiography*, 22.

6. The complete list of courses was Diatetics, Materia Medica and Pharmacy, by Dr. Duncan; Theory of Physics, Dr. Duncan; Clinical Lectures on Medicine, Dr. Alison; Clinical Lectures on Surgery, Mr. Russel; Military Surgery, Dr. Ballingall; Chemistry and Pharmacy, Dr. Hope; Anatomy, Physiology and Pathology, Dr. Munro; Principles and Practice of Surgery, Dr. Monro; Midwifery, Dr. Hamilton; and Practice of Medicine, Dr. Home. In the summer sessions, which Darwin did not attend, there were also offered Botany, Dr. Graham; Universal History, Sir William Hamilton; Clinical Lectures on Medicine, Dr. Home; Clinical Lectures on Surgery, Mr. Russel; and Medical Jurisprudence, Dr. Christian.

7. Darwin to Caroline, January 6, 1826, in *Correspondence*, 1:25.

8. DAR 5, Cambridge University Library, Cambridge.

9. John Struthers, *Historical Sketch of the Edinburgh Anatomical School* (Edinburgh: Machlachlan and Stewart, 1867), 38.

10. *Autobiography*, 23.

11. Darwin to Susan Darwin, January 26, 1826, in *Correspondence*, 1:28.

12. DAR 129, Cambridge University Library, Cambridge: diary for 1826, a small, red leather volume, purchased in Edinburgh.

13. Susan Darwin to Darwin, March 27, 1826, in *Correspondence,* 1:37.

14. Darwin to Caroline, April 8, 1826, in *Correspondence,* 1:39.

15. Susan Darwin to Darwin, March 27, 1826.

16. He seems to have shot for one week only. On September 1, he shot seven and a half brace of partridges and one hare. On the 2nd, six and a half brace of partridges; on the 3rd, two and half brace; on the 7th, three and a half brace plus a rabbit; on the 8th, three and a half brace plus one hare. After that there are no diary entries until October 28.

17. Janet Browne, *Charles Darwin, Voyaging* (New York: Knopf, 1995), 109.

FIVE

Robert Jameson

1. Some authors have stated that he signed up only for Home's course. The tickets for Darwin's courses in 1826–27 are preserved in DAR 5, Cambridge University Library, Cambridge.

2. *Autobiography,* 24.

3. DAR 5. Haggis is a surprisingly tasty dish consisting of various internal organs of a sheep (heart, liver, etc.) boiled with suet, oatmeal, and onion and spices in a sheep's stomach; ironically, its origin is English. Collops (derived from escalope) is a dish of veal, beef, or lamb slices, fried and then simmered in stock and served with bacon and mushrooms. The name is also used for a dish for more spartan budgets, of fried eggs and bacon. "Scotch" in this case does not mean Scottish but "scortched," or scored.

4. Josiah Wedgwood to Robert Darwin, August 31, 1831, in *Correspondence,* 1:134.

5. Offered under the division of Literature and Philosophy. Darwin's tickets for both the lecture course and study in Jameson's Natural History Museum are preserved in DAR 5. An anonymous article in *St. James's Gazette,* February 17, 1888, mistakenly stated that Darwin did not register for this class.

6. Georges Cuvier, *Récherches sur les ossemens fossiles de quadrupèds* (Paris, 1812); Cuvier, *Essay on the Theory of the Earth, with Mineralogical Notes, and an Account of Cuvier's Geological Discoveries by Professor Jameson,* first English ed., ed. Robert Jameson (Edinburgh: Blackwood, 1813).

7. Derek Flinn, "James Hutton and Robert Jameson," *Scottish Journal of Geology* 16 (1980): 251–58.

8. Cuvier, *Essay on the Theory of the Earth,* 57.

9. Ibid.

10. James Hutton, *Theory of the Earth* (1788).

11. *Autobiography,* 26.

12. Darwin to Adolf von Morlot, August 9, 1844, in *Correspondence,* 3:51.

13. James Secord, "The Discovery of a Vocation: Darwin's Early Geology," *British Journal for the History of Science* 24 (1991): 141.

14. Darwin to Fox, August 1, 1831, in *Correspondence*, 1:126.

15. Sir Alexander Grant, *The Story of the University of Edinburgh during Its First Three Hundred Years* (London: Longmans, Green , 1884), 2:433.

16. W. D. Wilson MS, Archives of the American Philosophical Society, Philadelphia.

17. Stewart's textbook is a very traditional tome, basically listing and characterizing animals in the manner begun by Linnaeus, defining six classes: Mammalia, Aves, Amphibia, Pisces, Insecta, Vermes. Most of his source material was published before 1780.

18. John Stark, *Elements of Natural History, Adapted to the Present State of the Science, Containing the Generic Characters of Nearly the Whole Animal Kingdom, and the Description of the Principal Species* (Edinburgh: Blackwood, 1828), 2:492.

19. Robert McCormick, Lectures on Natural Philosophy by Professor Jameson, MSS 3358, Wellcome Library, London.

20. *Autobiography*, 25.

21. This may be a reference to Scoresby's *An Account of the Arctic Regions* (1820). Scoresby's *Journal of a Voyage to the Northern Whale-Fishery* was not published until 1823, which is when Jameson reviewed it for the *Edinburgh Philosophical Journal.*

22. In 1827, a second "Account of the Capture of a Colossal Orang-Outan in the Island of Sumatra, and Description of Its Appearance" appeared in the *Edinburgh New Philosophical Journal* 3, 81. The animal, when pursued, "sought refuge in another tree at some distance, exhibiting as he moved, the appearance of a tall man-like figure, covered with shining brown hair, walking erect, with a waddling gait." In these accounts, there is a very great deal in common with Darwin's own observations of an orangutan at the Zoological Society of London in 1843: "Let man visit Ourang-outang in domestication, hear expressive whine, see its intelligence when spoken to: as if it understands every word said—see its affection" (Notebook C, 79).

23. One of Jameson's characteristics in this course, as for many contemporary scholars, was to see the exercise of classifying minerals and animals as essentially identical pursuits.

24. For 1830–31, Jameson divided the course into four sections: Meteorology, Hydrology, Geology, and Mineralogy and Zoology.

25. Some of the famous doodles on the inside cover look very much as though they were made by a child, presumably one of Darwin's, and therefore much later.

26. McCormick, in his diary for Saturday, April 2, 1832, wrote, "At two p.m. accompanied him [Jameson] and his class on an excursion to the Salisbury Crags—the geological structure and mode of formation of which, with the hills adjacent, he explained on the spot, and we returned at 3.30 p.m." On Saturday, April 9, McCormick "went with Professor Jameson and his class, on a geological excursion to the summit of Arthur's Seat; started at two, and returned at four." Quoted in McCormick, *Voyages of Discovery in the Arctic and Antarctic Seas, and Round the World* (London: Sampson, Low, 1884), 216.

27. *Autobiography*, 26.

28. Robert Jameson, "General Observations on the Former and Present Geological Condition of the Countries Discovered by Captains Parry and Ross," *Edinburgh New Philosophical Journal* 2 (October 1826): 104–7.

29. Jameson, in Cuvier, *Essay on the Theory of the Earth*, 189.

30. Stark, *Elements of Natural History*.

31. Grant, *The Story of the University of Edinburgh*, 433.

S I X

Mentors and Models

1. DAR 129, Cambridge University Library, Cambridge.

2. *Autobiography*, 23.

3. P. Helveg Jespersen, "Charles Darwin and Dr. Grant," *Lychnos* (1949): 159–67.

4. In 1830, Geoffroy and Cuvier had a confrontation in Paris over their opposing views: unity and plan and transformation (for the former), rigidity of adaptation and denial of change (for the latter). One of the most telling arguments for Cuvier's side was his principle of correlation of parts. He asserted that, even if one had only one or two bones from a skeleton, the lawlike relationship between structure and function predicted what the rest was like. With his ability to demonstrate this principle on fossils from the Paris basin, and having the advantage of dealing in practical rather than theoretical terms, Cuvier carried the day against the more evolutionary view; see Toby Appel, *The Cuvier-Geoffroy Debate: French Biology in the Decades Before Darwin* (New York: Oxford University Press, 1987).

5. DAR 129.

6. Keith S. Thomson and Stan Rachootin, "Turning Points in Darwin's Life," *Biological Journal of the Linnean Society of London* 17 (1982): 23–38.

7. *Autobiography*, 75.

8. Some authors have stated that Darwin did not take out a library ticket for 1826–27. The ticket is preserved in DAR 5, Cambridge University Library, Cambridge.

9. DAR .271.1:2, quoted by kind permission of Mrs. Ursula Mommens.

10. Ernst Krause, *Erasmus Darwin* (New York: Appleton, 1880).

11. William Lawrence, *Lectures on Physiology, Zoology, and the Natural History of Man* (London: Craddock, 1819), 2.

12. Ibid., 8.

13. This notebook is preserved as DR 118, Cambridge University Library. It contains notes on field and laboratory studies at Edinburgh and a small number of notes on botanical investigations made later in Cambridge. The Edinburgh notes seem to be an attempt to set down, after the fact and in a concise, mannered style, just what his contributions to the joint ventures with Grant had been. The work on

flustrae and the supposed egg masses of Fucus are written out in detail, as if for publication.

14. See facsimile in Adrian Desmond and James Moore, *Darwin: The Life of a Tormented Evolutionist* (New York: Norton, 1991), fig. 9.

15. Ibid.

16. DAR 118, Cambridge University Library, Cambridge.

17. Henrietta Litchfield, quoted in *Life and Letters of Charles Darwin*, ed. Francis Darwin (London: Murray, 1887), 1:130.

18. DAR 118.

19. Robert Grant, "Observations on the Structure and Nature of the Flustrae," *Edinburgh New Philosophical Journal* 3 (1827): 341.

20. Robert Grant, "Notice regarding the Ova of the *Pontobdella muricata*, Lam.," *Edinburgh Journal of Science* 7 (1827): 160–61.

21. Litchfield, quoted in Darwin, *Life and Letters of Charles Darwin*, 1:130.

22. *Autobiography*, 24.

SEVEN
Lamarckians

1. Sir Alexander Grant, *The Story of the University of Edinburgh during its First Three Hundred Years* (London: Longmans, Green, 1884). Interestingly, Robert McCormick, as a staunch churchgoing man, was unimpressed by Jameson's "philosophy." Normally his notes on Jameson's lectures were lengthy, but his record of the relevant lecture merely reads: "February 17th. Introductory Remarks on Zoology and on Life, and Organization. Distinguishing Characters of Animals. Structure—Elementary Form globular in all. Etc. etc. etc." (Robert McCormick, Lectures on Natural Philosophy by Professor Jameson, MSS 3358, Wellcome Library, London). By contrast, McCormick seems fairly to have reveled in the minutiae of mineral classification. Later Darwin was to observe caustically of him, "He was a philosopher of rather an *antient* date" (Darwin to Henslow, May 18, 1832, in *Correspondence*, 1:236–39.

2. Robert Hooke, *Discourse of Earthquakes*, 1694, in *The Posthumous Works of Robert Hooke, M.D., F.R.S.*, ed. Robert Waller (London: Royal Society, 1709), 291.

3. David Hume, *Dialogues concerning Natural Religion*, ed. Martin Bell (London: Penguin Classics, 1990), 53.

4. Georges-Louis Leclerc, comte de Buffon, *Histoire naturelle, générale et particulière*, 44 vols. (Paris: L'Imprimerie Royale, 1749–1809).

5. Hume, *Dialogues*, 88–89.

6. Georges Cuvier, *Essay on the Theory of the Earth, with Mineralogical Notes and an Account of Cuvier's Geological Discoveries by Professor Jameson*, first English ed. (Edinburgh: Blackwood, 1813); Cuvier, *Essay on the Theory of the Earth, with Geological Observations by Professor Jameson*, 5th ed. (Edinburgh: Blackwood, 1827).

7. *Autobiography*, 24.

8. Ibid.

9. Robert Grant, "On the Structure and Nature of the Spongilla Friabilis," *Edinburgh Philosophical Journal* 14 (1826): 270–84.

10. Frank N. Egerton, "Darwin's Early Reading of Lamarck," *Isis* 67 (1976): 452–56.

11. "Observations on the Nature and Importance of Geology," *Edinburgh New Philosophical Journal* 1 (October 1826): 293–302.

12. James Secord, "Edinburgh Lamarckians: Robert Jameson and Robert E. Grant," *Journal of the History of Biology* 24 (1991): 1–18.

13. "Of the Changes Which Life Has Experienced on the Globe," *Edinburgh New Philosophical Journal* 3 (1827): 288–300.

14. Darwin to Susan Darwin, September 4, 1831, in *Correspondence*, 1:139.The Humboldt reference was to the series of papers in the *Edinburgh Philosophical Journal* to which Jameson referred his students. Coldstream and Foggo together wrote papers on "Meteorological Observations Made at Leith" in 1825 and 1826. Coldstream separately wrote "Account of Some of the Rarer Atmospherical Phenomena Observed at Leith in 1825" for the same journal, and Foggo published his "Results of a Meteorological Journal Kept at Seringapatan" in the *Edinburgh Journal of Science* (1826).

15. Secord, "Edinburgh Lamarckians."

EIGHT

Cambridge Undergraduate

1. *Autobiography*, 26.

2. Darwin to Fox, December 24, 1828, in *Correspondence*, 1:71.

3. *Autobiography*, 29.

4. Ibid., 10. The placing of the first comma here is crucial!

5. Of the Darwin men in three generations, there were eventually at least five doctors, one vicar (great-uncle John Darwin), two lawyers, one army officer, and three who essentially became gentlemen scholars (Robert Waring Darwin, brother of old Dr. Erasmus Darwin, Sir Francis Galton, and Sir Francis Sacheverel Darwin).

6. *Autobiography*, 29.

7. Ibid.

8. Darwin to Erasmus, December 21, 1828, in *Correspondence*, 1:76.

9. Darwin to Fox, October 29, 1828, in *Correspondence*, 1:69–70.

10. Darwin to Fox, January 25–29, 1829, in *Correspondence*, 1:73.

11. Darwin to Fox, April 1, 1829, in *Correspondence*, 1:81. Charles Whitley had been at Shrewsbury School with Darwin. Another naturalist scholar, he later became reader in natural philosophy at Durham University and rector of Bedlington, Northumberland.

12. Ibid.

13. Darwin to Fox, February 26, 1829, in *Correspondence*, 1:75.

14. James Francis Stephens, *Illustrations of British Entomology* (London: Baldwin and Cradock, 1828–35).

15. Darwin to Fox, February 26, 1829,

16. *Autobiography*, 31.

17. Darwin to Fox, July 3, 1829, in *Correspondence*, 1:88.

18. Darwin to Fox, January 1830, in *Correspondence*, 1:98.

19. Darwin to Fox, July 15, 1829, in *Correspondence*, 1:89. In all, notices of thirty-five insects appeared in Stephens's work, and thirty-four of those were beetles. See *Correspondence*, 1:90n1.

20. *Autobiography*, 30.

21. Darwin to Fox, July 15, 1829, in *Correspondence*, 1:89.

22. *Correspondence*, 1:75n7.

23. Darwin to Fox, March 25, 1830, in *Correspondence*, 1:101.

24. Darwin to Fox, January 3, 1830, in *Correspondence*, 1:96.

25. Darwin to Fox, January 13, 1830, in *Correspondence*, 1:98.

26. *Autobiography*, 30.

27. Ibid., 31.

28. Ibid., 30.

29. John Graham, quoted in S. M. Walters and E. A. Stow, *Darwin's Mentor, John Stevens Henslow, 1796–1861* (Cambridge: Cambridge University Press, 2001), 79.

30. Darwin to Fox, November 5, 1830, in *Correspondence*, 1:109.

31. Darwin to Fox, November 27, 1830, in *Correspondence*, 1:111.

NINE
More Serious Things

1. *Autobiography*, 30.

2. William Sharp MacLeay, *Horae Entomologicae; or, Essays on the Annulose Animals* (London: Bagster, 1819–21).

3. William Swainson, *A Preliminary Discourse on the Study of Natural History* (London: Longmans, 1839), 190.

4. James Francis Stephens, *Illustrations of British Entomology* (London: Baldwin and Cradock, 1828–35), 1:2.

5. Darwin to Henslow, October 26, 1832, in *Correspondence*, 1:280.

6. One of Darwin's favorite books—Stephens's *A Systematic Catalogue of British Insects*—noted that some nineteen "kinds" of the beetle Coccinella variabilis were really only one species; Coccinella mutabilis had similarly been split into multiple species.

7. John Stevens Henslow, "On the Specific Identity of the Primrose, Oxlip, Cowslip, and Polyanthus," *Magazine of Natural History* 3 (1830): 406–9.

8. John Stevens Henslow, "On the Specific Identity of Anagalis Arvensis and Cerulean," *Magazine of Natural History* 3 (1830): 537–38. However, Henslow later discovered that the experimental sowings were suspect—the problem was the red-flowered plant, which turned up in the "control" plot in which no seeds had been placed. Nevertheless, Henslow insisted that "after a multitude of such experiments shall have been carefully performed, we can expect to arrive scientifically and legitimately at the truth." See *Magazine of Natural History* 5 (1832): 494.

9. Nicolaas A. Rupke, *The Great Chain of History: William Buckland and the English School of Geology, 1814–1849* (Oxford: Clarendon, 1983).

10. Whereas French scholars such as Elie de Beaumont had recently come down on the side of volcanoes as the source of the uplift of mountains, Lyell showed that elevation was a universal phenomenon, not restricted to mountains, and must, he thought, be due to earthquakes.

11. Arthur O. Lovejoy, *The Great Chain of Being: A Study of the History of an Idea* (Cambridge, Mass.: Harvard University Press, 1936).

12. Charles Lyell, *Principles of Geology,* vol. 1 (London: Murray, 1831), 123.

13. Adam Sedgwick, 1825, quoted in Anthony Hallam, *Great Geological Controversies,* 2nd ed. (Oxford: Oxford University Press, 1989), 43.

14. Adam Sedgwick, "Presidential Address," *Proceedings of the Geological Society of London* 1 (1831): 270–316.

15. In the late 1830s and early 1840s, the young Swiss scientist Louis Agassiz started to publish a theory that would revolutionize geology. The "water" that acted so powerfully had been ice. What appeared to be evidence of the Flood was, in fact, evidence of wide-scale glaciation in western Europe and beyond. One can analyze the mode of action of ancient glaciers by studying modern glaciers in the Alps. The evidence of rocks scarred with parallel lines from the scouring action of glaciers showed that ice had once covered a great deal of western Europe. The ice sheets had periodically grown and retreated, and the time frame involved was hundreds of thousands of years, at the very least. The "Diluvial" sand and gravel deposits were glacial deposits. Glaciation also explained the puzzling phenomenon of erratics—huge boulders of rock that could be found tens or hundreds of miles from their source. A Flood could not have transported such masses; ice could. Darwin was well familiar with one such erratic—the "Bell-stone" at Morris Hall in Shrewsbury. Legend had it that the world would end when someone could explain how it got there. And, in the intellectual sense, a world did end.

16. Eric Ashby and Mary Anderson, preface to Adam Sedgwick, *A Discourse of the Studies of the University of Cambridge* (Leicester: Leicester University Press, 1969), 16.

17. The book opens with a quotation from Psalms 116:16–19.

18. Sedgwick, *Discourse,* 24.

19. Ibid., 5th ed. (1850), 44.

20. Charles Lyell, "Presidential Address," *Quarterly Journal of the Geological Society of London* 7 (1851): xxv.

21. Sedgwick, *Discourse,* 23.

TEN
Reading Science

1. Darwin, *Journal,* in *Correspondence,* 1:539.

2. Darwin to John Lubbock, November 15, 1859, in *Correspondence,* 7:388.

3. Anonymous review of Paley, *Quarterly Review* 1 (1802–3): 287–95.

4. Keith Thomson, *Before Darwin: Reconciling God and Nature* (New Haven, Conn.: Yale University Press, 2005).

5. Kenneth R. Miller, "Falling over the Edge," *Nature* 447 (2007): 1055–56.

6. Thomas Robert Malthus, *An Essay on the Principle of Population* (London: Johnson, 1798).

7. William Paley, *Natural Theology; or, Evidences of the Existence and Attributes of the Deity* (London: Hallowell, 1801), 273. Paley's apology for poverty was not a mere theoretical argument in turn-of-the-century Britain and would become even more important as the century wore on. Poverty was a serious, draining social problem. The English Poor Laws have their origins in the reign of Queen Elizabeth the first, and ever since then arguments have been conducted in the same frame, one that is depressingly familiar to modern readers, particularly in the matter of what are called "entitlements." When someone is desperately poor, do you feed them and produce, thereby, more poor? More to the point, how do you prevent the poor from arming themselves and storming the capital?

8. Ibid., 37.

9. Ibid., 38.

10. Malcolm Nicholson, "Historical Introduction," in Alexander von Humboldt, *Personal Narrative,* ed. Malcolm Nicholson (London: Penguin Classics, 1995), ix.

ELEVEN
Geology Again

1. J. M. Herbert to Darwin, May 1831, in *Correspondence,* 1:122.

2. Darwin to Fox, April 7, 1831, in *Correspondence,* 1:120.

3. Humboldt, *Personal Narrative,* 29–30.

4. Darwin to Fox, July 9, 1831, in *Correspondence,* 1:124.

5. *Autobiography,* 31.

6. James Secord, "The Discovery of a Vocation: Darwin's Early Geology," *British Journal for the History of Science* 24 (1991): 133–57; Sandra Herbert, *Charles Darwin, Geologist* (Ithaca, N.Y.: Cornell University Press, 2005).

7. Michael B. Roberts, "Darwin at Llanymynech: The Evolution of a Geologist," *British Journal for the History of Science* 29 (1996): 469–78.

8. Darwin to Henslow, July 11, 1831, in *Correspondence,* 1:125. Secord, "Discovery," 144, suggested that Darwin had "concealed this abortive expedition from Henslow" because Darwin wrote to him only after the fact of trying his hand with

his new instruments. In fact, Darwin began his letter to Henslow with the apology and explanation that "I should have written to you sometime ago, only I was determined to wait for the clinometer." Darwin to Henslow, July 11, 1831.

9. After the work in the Vale of Clywd, Sedgwick may perhaps have taken Darwin on to the Isle of Angelsey.

10. A. P. Martin, *Life and Letters of the Rt. Honorable Robert Lowe, Viscount Sherbrooke* (London: Longsmans, Green, 1893), 19.

11. Sedgwick to Darwin, September 4, 1831, in *Correspondence*, 1:137–38.

12. Paul H. Barrett, "The Sedgwick-Darwin Geologic Tour of North Wales," *Proceedings of the American Philosophical Society* 118 (1974): 146–64.

TWELVE
HMS *Beagle*

1. Darwin to Henslow, July 11, 1831, in *Correspondence*, 1:126.

2. Darwin to Fox, August 1, 1831, in *Correspondence*, 1:127.

3. George Peacock to Henslow, August 6 or 13, 1831, in *Correspondence*, 1:127.

4. Henslow to Darwin, August 24, 1831, in *Correspondence*, 1:128–29.

5. Darwin to Susan Darwin, September 4, 1831, in *Correspondence*, 1:139.

6. Peacock to Darwin, August 26, 1831, in *Correspondence*, 1:129–30.

7. Darwin to Henslow, August 30, 1831, in *Correspondence*, 1:131.

8. Robert Darwin to Josiah Wedgwood, August 30–31, 1831, in *Correspondence*, 1:132.

9. Josiah Wedgwood to Robert Darwin, August 31, 1831, in *Correspondence*, 1:133–35.

10. George Barsala, "The Voyage of the Beagle without Darwin," *Mariner's Mirror* 59 (1956): 42–49; Keith Thomson, *HMS Beagle: The Story of Darwin's Ship* (New York: Norton, 1995).

11. One of the hostages, named Boat Memory, died of smallpox at Plymouth.

12. J. W. Gruber, "Who Was the Beagle's Naturalist?" *British Journal for the History of Science* 4 (1969): 266–82. See also H. L. Bursten, "If Darwin Wasn't the Beagle's Naturalist, Why Was He on Board?" *British Journal for the History of Science* 8 (1975): 62–69.

13. Thomson, *HMS Beagle*.

14. Francis Beaufort to Robert FitzRoy, September 1, 1831, in *Correspondence*, 1:135–36.

15. Darwin to Henslow, July 11, 1831, in *Correspondence*, 1:125.

16. Darwin to Susan Darwin, September 4, 1831, in *Correspondence*, 1:139.

17. Darwin to Susan Darwin, September 5, 1831, in *Correspondence*, 1:140–41.

18. *Beagle Diary*, 5.

19. Darwin to Henslow, October 30, 1831, in *Correspondence*, 1:176.

20. Darwin to Caroline Darwin, April 25–26, 1832, in *Correspondence*, 1:218–21.

21. Robert FitzRoy to Captain Beaufort, March 5, 1832; quoted in Richard Darwin Keynes, *The Beagle Record* (Cambridge: Cambridge University Press, 1979), 42.

22. *Beagle Diary*, 8.

23. Darwin to Henslow, October 30, 1831, in *Correspondence*, 1:176.

24. Thomson, *HMS Beagle.*

THIRTEEN

Epiphanies

1. Nora Barlow, "Robert FitzRoy and Charles Darwin," *Cornhill Magazine* 72 (1932): 493–510.

2. Darwin's diary states that they got the ship off by having the men run from one side to the other to rock it free. Surgeon McCormick's diary for the same date says that they were towed off by boats from HMS *Caledonia* (Special Collections NS 3359, Wellcome Library, London).

3. Darwin's copy still exists, battered and with many loose pages. One could scarcely imagine a more well-used book, although there are very few marginal annotations. This fits the patterns for Darwin's other books. In his early scholarly years, he did not annotate heavily. In *Principles,* he marked a few passages in the section on coral reefs, probably after returning to England.

4. *Autobiography,* 59.

5. Darwin to Robert Darwin, February 10, 1832, in *Correspondence,* 1:206.

6. *Beagle Diary,* 39.

7. Ibid., 39–40.

8. Ibid., 290–91.

9. Ibid., 327–28.

FOURTEEN

Storms and Floods

1. Alexander von Humboldt, *Personal Narrative,* ed. Malcolm Nicholson (London: Penguin Classics, 1995), 3.

2. Darwin to Henslow, May 18–June 16, 1832, in *Correspondence,* 1:236–38.

3. Darwin to Caroline Darwin, April 25–26, 1832, in *Correspondence,* 1:225–26.

4. Robert McCormick, *Voyages of Discovery in the Arctic and Antarctic Seas, and Round the World* (London, Sampson, Low, 1884), 222.

5. Darwin to Henslow, May 18–June 16, 1832, in *Correspondence,* 1:236–39.

6. FitzRoy, in Robert FitzRoy and Philip Parker King, *Narrative of the Surveying Voyages of HMS Adventure and HMS Beagle, 1826–1836,* 4 vols. (London, Henry Colburn, 1837–39), 2:56.

7. *Beagle Diary,* 35.

8. Darwin, "A Sketch of the Deposits Containing Extinct Mammalia in the Neighbourhood of the Plata," *Proceedings of the Geological Society of London* 2 (1837): 544.

9. Charles Lyell, *Principles of Geology,* vol. 2 (London: Murray, 1832), 545.

10. Humboldt, *Personal Narrative,* 6.

11. Lyell, *Principles of Geology,* 545.

12. Ibid., 562.

13. Ibid., 183.

14. T. H. Huxley: "The Reception of the *Origin of Species,*" in *Life and Letters of Charles Darwin,* ed. Francis Darwin (London: Murray, 1887), 2:187–97.

15. *Beagle Diary,* 118–19.

16. Darwin to Charles Whitley, July 23, 1834, in *Correspondence,* 1:396.

17. Fuegia was named for the basketlike coracle that some of the *Beagle*'s men had to use to rescue themselves when their boat was stolen (the deceased Boat Memory was named for the same event); Jemmy was named for the pearl button for which he was exchanged; York Minster was named for an eponymous rock formation.

18. *Beagle Diary,* 133.

19. Ibid., 136–37.

20. Ibid., 188.

21. FitzRoy, *Narrative,* 2:250.

22. Darwin to Catherine Darwin, May 22, 1833, in *Correspondence,* 1:311–14.

23. Darwin to Henslow, July 18, 1833, in *Correspondence,* 1:321–22.

24. Darwin to Catherine Darwin, April 6, 1834, in *Correspondence,* 1:379–82.

25. *Beagle Diary,* 216.

26. Ibid., 226.

27. Darwin to Henslow, July 24–27, 1834, in *Correspondence,* 1:394–96.

28. FitzRoy, *Narrative,* 2:657–58.

29. Darwin to Fox, May 23, 1833, in *Correspondence,* 1:315–17.

30. Darwin to Caroline Darwin, October 13, 1834, in *Correspondence,* 1:410–12.

31. FitzRoy to Francis Beaufort, September 26, in Richard D. Keynes, *The Beagle Record* (Cambridge: Cambridge University Press, 1979), 238.

32. Darwin to Henslow, November 8, 1834, in *Correspondence,* 1:420.

33. Ibid.

34. *Beagle Diary,* 276.

35. FitzRoy, *Narrative,* 2:413–14.

36. *Beagle Diary,* 334.

37. Nora Barlow, "Darwin's Ornithological Notes," *Bulletin of the British Museum (Natural History),* 2nd ser., no. 7 (1963): 85.

38. Ibid., 74.

39. Ibid. By masterly detective work, Frank Sulloway has demonstrated that this entry was probably made between June 18 and July 19, 1836. "Darwin's Conversion: The *Beagle* Voyage and Its Aftermath," *Journal of the History of Biology* 15 (1982): 325–96.

40. Barlow, "Darwin's Ornithological Notes," 73. Frank Sulloway says that Darwin added the observation on gradation of the bills "almost as an afterthought."

That is an odd conclusion, contradicted by Darwin's use of the linking word "moreover." "Darwin and His Finches: The Evolution of a Legend," *Journal of the History of Biology* 15 (1982): 7.

41. Darwin, *Journal and Remarks, 1830–1836*, 465. This book was first published in 1839 as a third volume (technically a supplement to the second volume) of Robert FitzRoy and Philip Parker King's *Narrative of the Surveying Voyages of HMS Adventure and HMS Beagle, 1826–1836*, 4 vols. (London: Henry Colburn, 1837–39). In this work, as well as documenting the second ("Darwin") voyage, FitzRoy completed the account of the *Beagle*'s first voyage (1826–30) led by Philip Parker King who, by this time, had emigrated to Australia. Darwin's *Journal and Remarks* was reissued separately in 1839 as *Journal of Researches into the Geology and Natural History of the Countries Visited during the Voyage of H.M.S. Beagle Round the World, under the Command of Capt. Fitz Roy, R.N* (London: Henry Colburn). Darwin prepared a second edition in 1845, as *Journal of Researches into the Natural History and Geology, etc.* By 1879 it had become *A Naturalist's Voyage Round the World* (London: Murray), and then in 1805 it acquired its present name, *The Voyage of the Beagle*.

42. FitzRoy, *Narrative*, 3:460.

FIFTEEN

First Thoughts on Evolution

1. Darwin to Henslow, October 30–31, 1836, in *Correspondence*, 1:512–15.

2. Darwin to Fox, November 6, 1836, in *Correspondence*, 1:517.

3. Adrian Desmond and James Moore, *Darwin: The Life of a Tortured Evolutionist* (London: Norton, 1991).

4. Darwin to Henslow, October 30, 1836, in *Correspondence*, 1:514.

5. Darwin to Jenyns, April 10, 1837, in *Correspondence*, 2:16.

6. Darwin to Lyell, August 9, 1838, in *Correspondence*, 2:96.

7. Lyell to Darwin, February 13, 1837, in *Correspondence*, 2:4.

8. *The Zoology of the Voyage of H.M.S. Beagle, under the Command of Captain Fitzroy, R.N., during the Years 1832 to 1836*, 5 vols., published with the approval of the Lords Commissioners of Her Majesty's Treasury, edited and superintended by Charles Darwin, Esq. (London: Smith, Elder, 1839–43).

9. Darwin, *Journal and Remarks*, preface.

10. Loren Eiseley, *Darwin's Century: Evolution and the Men Who Discovered It* (Garden City, N.Y.: Anchor, 1958).

11. Darwin to Henslow, November 4, 1837, in *Correspondence*, 2:54.

12. Darwin to Henslow, September 20, 1837, in *Correspondence*, 2:47.

13. See, for example, Frank Sulloway, "Darwin and His Finches: The Evolution of a Legend," *Journal of the History of Biology* 15 (1982): 1–53; Sulloway, "Darwin's Conversion: The *Beagle* Voyage and Its Aftermath," *Journal of the History of Biology* 15 (1982): 325–96; Martin Hodge, "Darwin Studies at Work: A Reevaluation of Three Decisive Years (1835–1837)," in *Nature, Experiment and the Sciences*, ed. T. H. Levere and W.R. Shea (Dordrecht: Kluwer, 1990).

14. Gould cavalierly dubbed the rhea as *R. darwini,* even though it had already been described by d'Orbigny as *R. pennata.*

15. John Gould, "Remarks on a Group of Ground Finches from Mr. Darwin's Collection," *Proceedings of the Zoological Society of London* 5 (1837): 4–7.

16. *Journal of Researches,* 1845 edition, 361. The phrase "mystery of mysteries" may have been taken from a letter from Herschel to Lyell, February 20, 1836.

17. Sydney Smith, "The Origin of the *Origin* as Discerned from Charles Darwin's Notebooks and His Annotations in the Books He Read between 1837 and 1844," *Advancement of Science* 16 (1960): 392.

18. *Charles Darwin's Notebooks, 1836–1844,* ed. Paul H. Barrett, Petĕr J. Gautry, Sandra Herbert, David Kohn, and Sydney Smith (Ithaca, N.Y.: Cornell University Press, 1987). It is quite impossible for any student of Darwin to give enough credit to the authors of the comprehensive edition of Darwin's notebooks, with their erudite annotations and explanations. Without this meticulous scholarship, the reader would be left floundering the dark.

19. One problem in reading the notebooks is Darwin's eccentric punctuation, particularly his use of the comma.

20. Notebook B, 47.

21. Ibid., 229.

22. An interesting difference between Buffon and Lamarck was that Buffon saw the Chain of Being as something built from the top (God, the angels, humans) down, while Lamarck saw it as extending up from the bottom (crystals, single-celled organisms) to the top.

23. Systems of taxonomy and classification have their own internal logic and entail bringing numbers of species together into "groups," like with like. If one opens any book, today, on the birds of South America, the pages devoted to dozens of strikingly similar tree creepers and hummingbirds instantly speak to this matter of "affinity" or "relationship." Similarity can also lead to a false appearance of affinity (analogy). The most striking example known today (but not to Darwin) is the existence of fossil saber-toothed "tigers," one kind being related to true carnivores, the other to marsupials like opossums and kangaroos.

24. Almost a century earlier, the great naturalist Linnaeus had modified the Creation story of Genesis into a scheme in which all animals and plants arose at once in a single small area of the Middle East, and then, as the earth changed and became more fertile, the offspring of these pioneers spread out to give the present, rather inconsistent, patterns of diversity.

25. *Beagle Diary,* 337.

26. Ibid., 383.

27. *Journal of Researches,* 1839 edition (London: Murray), 212.

28. Red Notebook, 113.

29. Ibid., 127.

30. Ibid., 130.

31. In *On the Origin of Species,* Darwin had shifted entirely to a gradual process of change, but he continued to use the "representative" terminology. "But it

may be urged that when several closely-allied species inhabit the same territory we surely ought to find at the present time many transitional forms. Let us take a simple case: in travelling from north to south over a continent, we generally meet at successive intervals with closely allied or representative species, evidently filling nearly the same place in the natural economy of the land. These representative species often meet and interlock; and as the one becomes rarer and rarer, the other becomes more and more frequent, till the one replaces the other. But if we compare these species where they intermingle, they are generally as absolutely distinct from each other in every detail of structure as are specimens taken from the metropolis inhabited by each. By my theory these allied species have descended from a common parent; and during the process of modification, each has become adapted to the conditions of life of its own region, and has supplanted and exterminated its original parent and all the transitional varieties between its past and present states. Hence we ought not to expect at the present time to meet with numerous transitional varieties in each region, though they must have existed there, and may be embedded there in a fossil condition. But in the intermediate region, having intermediate conditions of life, why do we not now find closely-linking intermediate varieties? This difficulty for a long time quite confounded me. But I think it can be in large part explained": Darwin, *On the Origin of Species by Means of Natural Selection, or the Preservation of Favoured Races in the Struggle for Life* (London: Murray, 1859): 173–74.

32. *Journal of Researches*, 210.

<h2 style="text-align:center">SIXTEEN
Notebook B</h2>

1. Journal, June 26, 1837, published in *Correspondence*, 2:431.

2. Red Notebook, 7.

3. Trying to make sense of Darwin's notebook entries as he jumps from subject to subject, constantly referring to contextual thoughts that he has failed to write down, can drive the expert to distraction. As a kindness to the reader, only a small selection of the entries is quoted here.

4. Notebook B, 173n12–1 (by David Kohn). Subsequent page references to Notebook B in this chapter will be inserted parenthetically in the text.

5. In his highly influential *Evolution: The History of an Idea*, Peter Bowler (Berkeley: University of California Press, 1984) stated that this was "*not* meant to be a system of genealogical relationships: no one part of the chain is derived from any other" (his emphasis). However, Lamarck (*Philosophie zoologique,* trans. Hugh Elliot Whoops [Chicago: University of Chicago Press, 1984], 176) makes clear that it is such a representation: "We cannot doubt that the reptiles, by means of two distinct branches, caused by the environment, have given rise, on the one hand, to the formation of birds and, on the other hand, to the amphibian mammals, which have in their turn given rise to all the other mammals."

6. Martin Barry, "Further Observations on the Unity of Structure in the Animal Kingdom," *Edinburgh New Philosophical Journal* 22 (1837): 346. After this was published, Darwin took out a subscription to the journal.

7. Owen, who eventually came to be a rival of Darwin and never accepted the evolutionary view, took the field of theoretical morphology and tried to develop a theoretical position opposite to Darwin's; see Nicolaas A. Rupke, *Richard Owen, Victorian Naturalist* (New Haven, Conn.: Yale University Press, 1994); Adrian Desmond, *Archetypes and Ancestors* (Chicago: University of Chicago Press, 1984).

8. Notebook D, 35; see Richard Owen, "On the Anatomy of the Southern Apteryx," *Transactions of the Zoological Society of London* 2 (1841): 249.

9. Notebook E, 89.

10. Etienne Geoffroy Saint-Hilaire, *Principes de philosophie zoologique* (Paris, 1830).

11. Barry, "Further Observations," 126–27.

12. Discussions in Robert J. Richards, *The Meaning of Evolution* (Chicago: University of Chicago Press, 1992).

13. Phillip Sloan has pointed out that, if there were these three main groupings, but each of them had representatives in the other environments, there would be a system in fives: "The Making of a Philosophical Naturalist," in *The Cambridge Companion to Darwin,* ed. Jonathan Hodge and Gregory Radick (Cambridge: Cambridge University Press, 2003), 17–39.

14. H. E. Gruber (*Darwin on Man* [New York: Dutton, 1974]) states that Darwin's interest in monads was only fleeting. The number of notebook entries referring directly or indirectly to monadism is great, however, and they explain the references to the subject in the first edition of the *Journal of Researches* (212).

15. Lamarck, *Philosophie zoologique,* 238.

16. Kohn (Notebook B, n23–1) points out that in *Journal of Researches,* Darwin noted that this was an idea of Lyell's.

SEVENTEEN
Moving Forward, Living a Lie

1. Darwin to Henslow, September 20, 1837, in *Correspondence,* 2:48.

2. Darwin to Henslow, October 14, 1837, in *Correspondence,* 2:50–52.

3. Darwin to Caroline Wedgwood, May 1838, in *Correspondence,* 2:85.

4. Darwin to Fox, June 15, 1838, in *Correspondence,* 2:91.

5. Darwin to Lyell, August 9, 1838, in *Correspondence,* 2:96.

6. Ibid.

7. *Autobiography,* 96.

8. Glen Roy Notebook, 3–4.

9. Darwin to Lyell, August 9, 1838.

10. Darwin, personal journal, September 6, 1838, in *Correspondence,* 2:432. See also C. R. Darwin, "Observations on the Parallel Roads of Glen Roy, and of

Other Parts of Lochaber in Scotland, with an Attempt to Prove That They Are of Marine Origin," *Philosophical Transactions of the Royal Society* 29 (1839): 39–81.

11. A full transcript of Darwin's notes on possible marriage is given in *Correspondence*, 2:443–45.

12. Though it is true that there is no treasure trove of Darwin jokes, passed down through the family. Hilarious stories beginning with "Do you remember the time that Charles said . . ."do not exist.

13. At the end of Notebook C, Darwin gives a list of works he has read and one of those that he must read. The total is 174.

14. Keith S. Thomson, "Natural Science in the 1830s," *American Scientist* 74 (1986): 397–99.

15. At just this time (May 1838), Darwin went to Cambridge to read MacLeay's *Horae Entomologicae,* and he and Covington copied out passages. *Correspondence*, 2:87n1. Darwin's dinner with MacLeay at the Athenaeum was in August.

16. Notebook C, 63.

17. Darwin to Susan Darwin, April 1, 1838, in *Correspondence*, 2:80.

18. Notebook C, 79.

19. Notebook B, 34.

20. Notebook C, 30.

21. Ibid., 175.

22. *Autobiography,* 72.

23. Notebook D, 23.

24. Ibid., 31.

25. Ibid., 36.

26. Ibid., 58.

27. Ibid., 69.

28. Ibid.

29. Ibid., 51.

30. *Autobiography,* 72.

31. Darwin, personal journal, in *Correspondence*, 2:432.

32. Keith Thomson, *Before Darwin: Reconciling God and Nature* (New Haven, Conn.: Yale University Press, 2005), chapters 11–12.

33. Notebook B, 37.

34. In the *Autobiography,* 72, Darwin said that he "happened to read [Malthus] for amusement." Whether this can be taken at face value is hard to know, especially given his prior knowledge of Malthus from Paley's *Natural Theology.* The list of books that Darwin read in October (Notebook C) does contain a few references to works that he might have read "for amusement," but the suspicion that he might have been revisiting old reading is reinforced by the fact that at the same time he reread Herschel's *Preliminary Discourse.*

35. Notebook D, 134–35.

36. Ibid., 135.

37. Augustin Pyramus de Candolle, "Géographie botanique," in F. H. Cuvier, *Dictionnaire des sciences naturelles* (Paris: F.G. Levrault, 1820).

38. Notebook E, 3.

39. Ibid., 35.

40. Ibid., 71.

41. Ibid., 118.

42. David Hume, *The Philosophical Works of David Hume,* 4 vols. (Edinburgh: Adam Black, 1826). The references are to *Dialogues concerning Natural Religion* and Hume's essay *Of the Sceptical and Other Systems of Philosophy.* See Darwin, Notebook N, 101.

43. *Autobiography,* 72.

44. Notebook, C, 177.

45. Darwin to Emma, August 7, 1838, in *Correspondence,*2:95. See n2.

EIGHTEEN
Finding His Place

1. Darwin to Jenyns, June 24, 1841, in *Correspondence,* 2:292–93.

2. Maria Edgeworth, December 1840, in *Correspondence,* 2:255n2.

3. This led Anthony Campbell and Stephanie Mathews to suggest that Darwin suffered from lactose intolerance: "Darwin's Illness Revealed," *Postgraduate Medical Journal* 81 (2005): 248–51.

4. The most comprehensive treatment of Darwin's illnesses is that of Robert Colp, *To Be an Invalid* (Chicago: University of Chicago Press, 1977); see also George Pickering, *Creative Malady* (Oxford: Oxford University Press, 1974), who emphasizes the psychosomatic side.

5. Ferando Orrego and Carlos Quintana, "Darwin's Illness: A Final Diagnosis," *Notes and Records of the Royal Society of London* 61 (2007): 23–29.

6. Darwin to Lyell July 6, 1841, in *Correspondence,* 2:297–98.

7. Darwin to Lyell, September 6, 1861, in *Correspondence,* 9:256–57.

8. Darwin to Fox, September 4, 1843, in *Correspondence,* 2:387.

9. Darwin to FitzRoy, October 1, 1846, in *Correspondence,* 3:345.

10. Randal Keynes, *Annie's Box: Charles Darwin, His Daughter and Human Evolution* (London: Fourth Estate, 2001).

11. *Autobiography,* 72–73.

12. Darwin to Fox, June 15, 1838, in *Correspondence,* 2:91.

13. Darwin to Lyell, September 14, 1838, in *Correspondence,* 2:104–7.

14. Darwin to Henslow, November 1839, in *Correspondence,* 2:237–38.

15. Darwin to Fox, January 25, 1841, in *Correspondence,* 2:278–79.

NINETEEN
First Drafts

1. Francis Darwin, *The Foundation of the Origin of Species* (Cambridge: Cambridge University Press, 1909). All quotations used in this chapter are taken

from the newly edited online version of the 1842 *Sketch* at Darwin-online.org.uk, with the editing marks mostly removed for clarity.

2. Darwin, *Foundation*, xxi.

3. Notebook C, 176.

4. The reference to a ship, as an example of a highly complex structure, may have come from Hume or from Mandeville's *Fable of the Bees*. Or it may just have been his own idea. See Stephen G. Alter, "Mandeville's Ship: Theistic Design and Philosophical History in Charles Darwin's Vision of Natural Selection," *Journal of the History of Ideas* 39 (2008): 461–65.

5. Darwin to J. D. Hooker, January 11, 1844, in *Correspondence*, 3:1–2. Walter Cannon, in his edition of Herschel's famous "mystery of mysteries" letter to Lyell, states that the "murder" Darwin confessed to was a "break with the Uniformitarianism of his friend and mentor Charles Lyell" (*Proceedings of the American Philosophical Society* 105 [1961]: 302). Darwin's meaning, however, is perfectly clear: the "murder" was the death of the immutability of species. I find Cannon equally unconvincing in his assertion, in the same article, that Darwin "was able to be almost completely insensitive to theological considerations concerning the origin of species."

6. Darwin to Emma, July 5, 1844, in *Correspondence*, 3:43–44.

7. *Autobiography*, 73. Francis Darwin pointed out in his edition of the 1842 *Sketch* and 1844 *Essay* that divergence is explicit in several places there.

TWENTY
Crisis and Resolution

1. Darwin to Lyell, May 3, 1856, in *Correspondence*, 6:99–101.

2. *Autobiography*, 75.

3. Darwin to Asa Gray, September 15, 1857, in *Correspondence*, 6:447–49.

4. *Autobiography*, 73.

5. Charles Darwin and Alfred Russel Wallace, "On the Tendency of Species to Form Varieties; and on the Perpetuation of Varieties and Species by Natural Means of Selection," *Journal of the Linnean Society of London* 3 (1858): 53–62.

Index

Note: Page numbers in italic type indicate illustrations.

267